Computer
Numerical Control
Accessory Devices

Other McGraw-Hill Books by Mike Lynch

Computer Numerical Control for Machining
Computer Numerical Control Advanced Techniques

Computer
Numerical Control
Accessory Devices

Mike Lynch

McGraw-Hill, Inc.

New York   San Francisco   Washington, D.C.   Auckland   Bogotá
Caracas   Lisbon   London   Madrid   Mexico City   Milan
Montreal   New Delhi   San Juan   Singapore
Sydney   Tokyo   Toronto

Library of Congress Cataloging-in-Publication Data

Lynch, Mike.
    Computer numerical control accessory devices / Mike Lynch.
        p.   cm.
    Includes index.
    ISBN 0-07-039226-9
    1. Machine-tools—Numerical control.   I. Title.
TJ1189. L94   1993
    621. 9'023—dc20   93-8902
                                                                           CIP

1 2 3 4 5 6 7 8 9 0   DOC/DOC   9 9 8 7 6 5 4 3

ISBN 0-07-039226-9

*The sponsoring editor for this book was Robert W. Hauserman, the
editing supervisor was Mitsy Kovacs, and the production supervisor
was Pamela A. Pelton. This book was set in Century Schoolbook by
McGraw-Hill's Professional Book Group composition unit.*

Printed and bound by R. R. Donnelley & Sons Company.

Dedicated to the memory of my father,
Leonard J. Lynch

Contents

Preface

Computer numerical control (CNC) machines are currently involved with almost all facets of manufacturing. There is no manufacturing industry not touched in some way by what these very productive and cost-effective machine tools can do.

Machine-tool builders, in order to supply the most efficient and distinctive CNC machine tools, supply a wide spectrum of special accessories designed to enhance or extend what the CNC machine tool is capable of doing. Also, the diversity of CNC equipment has made it possible and even attractive for companies related to CNC to parallel what the machine-tool builders are doing in the way of accessory devices. Just as the machine-tool builders do, many peripheral companies now supply aftermarket devices to enhance the performance and/or capabilities of this sophisticated kind of equipment.

No longer is the CNC machine tool the only factor contributing to the success of the CNC environment. In fact, in many cases, the success of the CNC environment depends heavily on the proper application of accessory devices.

The evolution of aftermarket accessory devices for use with CNC can be easily compared to the recent revolution in personal computers (PCs). The personal computer by itself is of little value without software and certain other accessories. Along with computer manufacturers, a number of software suppliers have been successful in producing a wide range of software products to increase the number of applications for personal computers.

Software is but one of the aftermarket accessories that can be purchased for today's personal computers. Hardware devices like the mouse, hand scanners, full-page scanners, modems, and printers of all varieties are but a few examples of aftermarket devices that can dramatically increase the personal computer's applications and capabilities.

In similar fashion, there are a number of accessories available for CNC equipment. As with the personal computer, some are supplied and supported completely by the machine-tool builder or control manufacturer. But, by far, a greater number are available from aftermarket suppliers that have little or no relationship with the machine-tool manufacturing companies.

As with the personal computer example, aftermarket suppliers for CNC equipment have a greater motivation to make theirs the products that perform best. Their very existence depends on how well they do in this regard.

The Lack of Emphasis on Accessory Devices

Since CNC machines by themselves do make up such a large portion of the CNC environment, often a CNC beginner is so concerned with becoming proficient with programming or operation of the machine tool itself that certain important factors of the CNC environment are left ignored and unstudied. Granted, the proper application of the machine tool itself plays a major role in the success of the company. However, it is but one facet of the whole CNC picture. Learning CNC without including a study of accessory devices would be like trying to learn about computers without considering software, printers, and other computer accessories. Without a good understanding of those accessories required for use with CNC, the CNC person cannot hope to take full advantage of all that is possible with the equipment.

Often the proper application of accessory devices makes the difference in a CNC machine's success or failure. For example, almost all companies utilizing CNC equipment use some form of program preparation device. This program preparation device may take the form of a simple text editor or a computer-aided manufacturing (CAM) system. To make the most of the entire CNC environment, the user must not only master the usage of the CNC machine tool itself, but must also master the program preparation device as well. Without a good understanding of this accessory device, programs for the CNC machine cannot even be created!

This is but one simple example of how accessory devices impact on the success of the CNC environment. In *all* cases, the CNC user must first be able to recognize that a certain accessory is available, and will help in the particular CNC environment. Once recognized, the user must be able to make the best use of the device.

This is a common scenario in many fields. A race car driver must possess a high level of driving expertise as a primary concern. However, the more the driver knows about internal-combustion engines, fuels, braking systems, suspension systems, tires, and all other facets of the automobile, the better the driver can be. Musicians must be most competent with the instruments they play. But the more they know about other instruments, composition, and playing style, the better they will be. The marksman may be most concerned with aiming and firing a gun. But the more he or she knows about gun maintenance, load styles, and sight adjustments, the better the marksman can be.

In the same way, a CNC user working with any form of CNC equipment must first possess a firm and complete knowledge of the machine tool itself. But just as importantly, the user must master every accessory device included in the CNC environment.

For example, for working with a machining center, a wide range of accessories may be available. The machining center user may have to master the use of a pallet changer, a probing system, an automatic tool changer, a special work-holding system (like fixtures of all kinds), tooling, and program preparation systems in order to be able to make the most of the machine tool itself. In like manner, a turning center user may have to master the use of a bar feeder, tailstock, steady rest, and live tooling. The list of potential accessory device applications goes on and on.

Unfortunately, there are limited places to which a CNC person can turn to learn about accessory devices for CNC. While most machine-tool builders offer training for the machine tool itself, most assume the user will independently figure out any accessory device equipped with the machine. While they may be able to answer specific questions about any one device, no formal training is usually scheduled for accessories. Worse, if the accessory device is supplied by an aftermarket manufacturer, the machine-tool builder may refuse to help at all.

The supplier of each accessory device may be able to help with the specific usage of its particular device, but unwilling or unable to discuss how the device is interfaced with the machine tool itself.

In both situations (machine-tool builder and aftermarket supplier), if training is available, it would not even begin until *after* an accessory device is purchased. Most programmers exposed to this kind of learning environment would agree that it is a baptism-by-fire approach to learning.

Most technical schools offer CNC training, but like machine-tool builders, their primary concern will be to present the usage of the CNC machine tool itself. Very few curriculums offer any presentations related to accessory devices.

How This Book Can Help

This book is unique in the sense that its *only* goal is to present information about accessory devices related to CNC. There are several ways in which it can help anyone attemping to learn more about CNC.

Exposure to a multitude of different devices

There is a saying that applies to all forms of learning: "Before you can begin to apply any technique or feature, you have to know that the technique or feature is available!" For example, if your telephone has an automatic redial feature, you have to know what the feature is *and* that it is equipped on your phone *before* you can even attempt to learn how to use the feature.

The field of CNC is filled with special accessories aimed at making a CNC machine perform better. But before a CNC user can begin to take advantage of these accessories, he or she must know what they are and what they can do to improve the CNC environment.

This book will acquaint the reader with countless accessories and techniques available for CNC. While not every type of accessory will be of immediate concern, at least you will have been exposed to the accessory for the time when the need arises.

Learn about CNC accessory devices without pressure

Nothing is worse than having to learn something new while under a great deal of pressure. If you wait until the day that you have to actually work with one accessory or another to begin learning about the accessory, you will be under a great deal of pressure. Possibly, *you* will be the bottleneck that keeps production from being run. It is very difficult to learn anything new while you have production people breathing down your neck, wondering why you cannot make the CNC machine and/or accessory device function.

By using this book with which to become familiar with potential future needs, you will be much better prepared to work with the accessory device when the need arises. While there are numerous variations in the way certain accessory devices function, at the very least you will have been exposed to the most basic and common possibilities.

Learn specific application and programming techniques

Most machine-tool related accessory devices discussed in this book require programming considerations. While machine-tool and accessory-device manufacturers do vary somewhat with regard to how certain devices are applied and programmed, this book will show the most common methods. In most cases, specific programming examples will be shown, to stress the points being made. Armed with the reasoning behind how a particular device functions *and* with the specific programming techniques related to one specific device, you should be well prepared for any variations that come along.

Learn to match your requirements with your company's budget

As with almost any facet of manufacturing, CNC and related accessory devices sometimes require compromises to be made. Just as a company may not be able to justify the purchase of an expensive single-purpose CNC machine tool and compromises must be made by using general-purpose machines, sometimes compromises have to be made with the purchase of accessory devices.

When applicable, we will show possible alternatives and list the compromises and limitations that the CNC user can expect. While our presentations may get a little opinionated during these discussions, at least you'll know what you will be in for if compromises need to be made.

Organization

We think you'll find that this book can easily be used in two different ways. The first is as a tutorial. A relative newcomer to CNC should be able to read the book from cover to cover and find that the presentations are made in a logical and easy to follow manner. By reading the book in this manner, you can gain a wealth of new information related to CNC.

However, there will be those readers who have a specific and imperative need to learn about only one particular accessory device. Such readers can use the table of contents and index to find the information they seek. Once the reader has been directed to the proper section, the presentation will be self-contained and will provide the necessary background required to understand the information presented.

This book is presented in three rather lengthy chapters. Chapter 1 presents CNC-related devices used in the preparation, transfer, storage, and verification of CNC programs. As you will see, there are numerous devices available for use in this area. This chapter will explore the application and usage for these important CNC devices.

Chapter 2 presents machining-center accessory devices. We start with simple devices and work toward more complex ones. Numerous devices, as well as specific programming techniques required for each, will be presented.

Chapter 3 switches the focus to turning-center accessory devices. While certain devices are common to both turning and machining centers, most are applied quite differently from one style of machine to the other. In most cases, presentations made in this chapter will be complete unto themselves, discussing only how the device is applied to turning centers.

Author's Note

Admittedly, no text can thoroughly cover every minute detail of how each CNC accessory is justified, applied, and programmed. The variations within each device category make such a goal impossible. Truly, there are certain accessories (like probing devices and automation systems) that could fill a volume by themselves. Our intention is *not* to become bogged down trying to present every little detail and restriction related to CNC accessories. Instead, our intention is to acquaint you with each accessory, giving you a working knowledge of how it is applied and programmed.

The author welcomes comments and suggestions related to the current text, as well as ideas for devices which the reader feels should be added for future editions.

Acknowledgments

For the photos used throughout this text, the author would like to thank the following companies:

CNC Concepts, Inc.

Fanuc U.S.A. Corporation

GE Fanuc Automation

Mckenna Service Company

Mori Seike U.S.A.

Okuma & Howa Machinery U.S.A.

Softwerks International

Thanks also go to Chris De Hut, Pete Nagel, and Mark Davis for their help with content suggestions and criticisms.

Mike Lynch

1

Program Transfer, Preparation, and Verification Devices

Every company utilizing computer numerical control (CNC) equipment is highly concerned with being able to efficiently create, maintain, store, retrieve, and verify CNC programs. Their degree of concern in this regard depends on many factors, including the type of company, the number of CNC machines in the company, the complexity of the work, the number of different workpieces to be machined, and how often new setups are made. These sometimes conflicting factors can make it difficult to choose which program preparation, storage/retrieval, and program verification systems to use.

Generally speaking, product manufacturing companies that machine a limited number of different workpieces repeatedly may not be very concerned with program preparation devices since they run the same programs over and over again. Once their limited number of programs is prepared, they will have little need to create new programs. This kind of company will be much more concerned with the efficient transfer of existing programs to the machine.

On the other hand, companies that run a vast number of different workpieces in small quantities (like many job shops) are very concerned with being able to prepare programs efficiently. This kind of company will do anything in its power to minimize the time and effort it takes to prepare and transfer programs.

The number of workers available also plays an important role in a company's program preparation and storage/retrieval device choice. In large manufacturing companies, the emphasis is placed on production. These companies wish to keep their CNC machine tools in production as much as possible. They will go to great lengths to perform *all* functions possible off line (away from the CNC machine). To achieve this, they employ several people in their CNC environment. One person may prepare the program (off line). Another may set all tools. Yet another will make the work-holding setup. And so on.

Ideally, this kind of company wants the CNC machine operator's responsibility to be limited to simply running production. The kind of program preparation and storage/retrieval devices chosen by this kind of company must suit the company's priorities in this regard. To have the operator entering programs at the machine (shop floor programming) in this kind of environment would be considered by most manufacturing people to be a waste of the machine's production time.

Conversely, smaller companies that run lower production quantities (like many job shops) cannot afford to hire a great number of people to support their CNC machines. In this kind of an environment, the CNC machine operator may be expected to do much more than simply run production. The operator may be required to program, set up, and verify all programs run on the machine as well as run production. In this situation, shop floor programming done by the CNC machine operator may make more sense than an off-line programming system.

The complexity of the work also determines what the company will need in the line of program preparation devices. Generally speaking, the more complex the work, the more help the programmer will need. For simple work that can be easily accomplished by manual programming techniques, the company may elect to use nothing more than a CNC text editor. The programmer will simply type the manual program exactly as it will be entered into the CNC control. On the other hand, if the program requires many difficult calculations to be made, the programmer may need the help of a more elaborate CNC program preparation system.

Organization of This Chapter

As you are beginning to see, there are many factors which influence a company's selection of form/s of accessory devices for program preparation. In this chapter we will look at many CNC program transfer, preparation, and verification devices. We will also show the actual methods used to create, store/retrieve, and verify CNC programs. We will look at these methods in order of complexity, working from simple to complex.

As you read this chapter, keep in mind that there are many variations and considerations related to the basic points we make. There is a great deal of disagreement and confusion that exists in industry as to which methods are best for any one company. While we will give our opinions in this regard, our most basic goal is to acquaint you with the devices available. And while our opinions should steer the beginner in the correct direction, experienced CNC people may not totally agree with every point we make. Regardless of your company's current or future situation, when you finish this chapter you should be able to make informed decisions as to which devices are best for your particular CNC environment when the need arises.

We will first introduce the most common devices used to transfer programs from one CNC device to another. You will find that program transfer devices are used to transfer programs to and from other devices addressed in this chapter.

Second, we will look at the most common forms of program preparation devices, showing you numerous methods companies use to help prepare programs. These devices include the CNC control itself, CNC text editors, and computer-aided manufacturing (CAM) systems.

Third, we will introduce the two forms of DNC, distributed numerical control and direct numerical control. There is much confusion in the industry regarding these two CNC functions. We hope to clarify this confusion.

Finally, we will discuss program verification devices, showing you how many companies attempt to verify their CNC programs with the help of devices other than the CNC machine tool itself.

Program Transfer Devices Used in CNC Shops

Once a CNC program is created, there are many ways to transfer the program to and from the CNC machine tool. There are also several ways to keep (store and retrieve) the program for future use. Before we discuss program preparation devices, we want you to understand the methods of program transfer they use. Though at this time we will do little more than define each device, at least you will have a basic understanding of what each program transfer device is as we discuss how programs are actually transferred a little later.

Tape reader/punches

One of the oldest (and still somewhat common) forms of program transfer is 1-in-wide paper tape. Though this form of program medium is becoming less and less common, you may still encounter it in your CNC environment. With 1-in-wide paper tape, a series of holes is punched in a row to represent one character of the CNC program. One character is punched per row of the tape; the tape includes many hundreds of rows of holes (characters), which taken together, make up the various words and commands of the CNC program.

For shops that have only relatively new CNC equipment (under 5 years old) it is unlikely that paper tape is being used as the program transfer medium. In fact, many CNC controls manufactured today do not even have tape readers. But for shops that have very old CNC (or NC) machines (over 15 years old), it is likely that 1-in-wide paper tape is the *only* way to transfer programs to the machine tool. Since many shops have a combination of very new as well as very old CNC (or NC) machine tools, they must still support at least some of their machines with 1-in-wide paper tape.

To transfer the program (on tape) to the CNC machine tool, a tape reader must be used. In most cases, the tape reader itself is incorporated within the CNC control (especially on older models). The CNC machine operator simply loads the tape into the tape reader and makes the command to make the control read the tape. As the tape passes through the reader, the holes punched in coded form transfer the commands included within the CNC program to the CNC control. When reading is finished, the program will be contained within the CNC control's memory.

Once the program resides in the memory of the CNC control, it will be available for modification. During the program's verification, it is likely that some changes will have to be made. If nothing else, cutting-condition changes (to speeds and feeds) will probably be necessary to optimize the program.

If changes are made to the CNC program within the control's memory, the program in the CNC control will no longer match the program that was loaded from the tape. If the company will need to machine the workpiece again at some later date, a new, corrected version of the tape must be made. For this reason, most companies using paper tape as their program transfer medium will utilize a portable tape punch. Once the program is verified and running production, they will connect the portable tape punch to the CNC control and punch the corrected version of the program from the control's memory to store it for future use.

Some portable tape punches also have the ability to read tapes as well. In this case, a tape reader head is incorporated into the punching device. When the tape device can both read and punch tapes, it is called a *tape reader/punch*. Tape reader/punches allow paper tape to be used to transfer programs among devices that do not allow paper tape. Figure 1.1 shows an example of a reader punch.

Here is one common situation that requires the use of a reader/punch. Say a company has just purchased a brand new CNC machining center that does not have a tape reader built into the con-

Figure 1.1 Portable reader/punch used to transfer programs to and from the CNC control utilizing one-inch-wide paper tape. (*Courtesy GE Fanuc Automation.*)

trol. Say the company also has several older CNC (or even older NC) machining centers that use paper tape as their only program transfer medium. At some point, the company will probably need to load an existing program from tape into the new CNC machining center. To do this will require the use of a portable tape reader/punch. (Note that portable tape readers do exist without the punching capability, but they are less common than tape reader/punches.)

Tape reader/punches are often required for use with program preparation systems. They allow programs to be created by the program preparation device to be used with CNC (or NC) machine tools that require paper tape as the program transfer media.

A note on two-way program transfer. Though this may be a little off the subject at hand, we want to make an important point about *any* program transfer device. To allow two-way program transfer (to *and* from the reader/punch, for example) requires that both communicating devices have a certain compatibility with regard to communications. All current forms of CNC equipment employ a standard of communications called an *RS-232C* or *serial* interface. While other communications protocol can also be used, by far the most common protocol for CNC usage is RS-232C. Both communicating devices must adhere to this standard. While this may sound a little complicated, all we are trying to say is that any two devices (reader/punch, CNC control, or any other device) adhering to this communications standard can communicate with each other. And any device that is required to both send *and* receive CNC programs must have this form of serial interface.

Some (especially older) portable tape punches employ a different standard for communications called a *parallel interface*. This standard is commonly used with printers on personal computers. A parallel interface allows only one-way communication, meaning a parallel-style tape punch can only punch tapes. There is no such thing as a parallel tape reader/punch that can both read and punch programs unless it has *both* a serial and parallel interface (as some do). Parallel-style portable tape punches were employed heavily in the early days of CNC (in the late 1970s and early 1980s) and were used strictly to punch programs in tape form from the CNC machine.

Portable floppy-disk drive systems

With the growing popularity of personal computers and serial communications, paper tape is fast becoming a thing of the past. Generally speaking, the only companies that still use paper tape to any extent are those that must because they have older CNC (or NC) machines that allow no other method of program transfer.

Paper tapes are cumbersome to work with, require large storage containers, and are quite difficult to catalog and store. About the only advantage of paper tape in today's technology is the fact that they are physically "solid." By this we mean that they cannot be affected by electrical surges and magnetic fields the way computer-related program transfer devices can. If they do become damaged, the damage is visible. This cannot be said for most other forms of program transfer media.

Since paper tape is so cumbersome to work with, more and more companies are finding other ways to store and retrieve programs. Computer floppy diskettes make an excellent program storage and transfer medium, requiring only a fraction of the physical storage space of paper tapes. For example, it is likely that programs requiring a whole file cabinet drawer full of paper tapes can be stored on one $3\frac{1}{2}$-in floppy diskette!

For this reason, more and more CNC control manufacturers are incorporating floppy-disk drives within the control itself, making it easy to store massive amounts of CNC program data. Once a CNC program is loaded into the control's memory (either through the control's keyboard or from some outside device), it can be easily copied to the floppy-disk drive of the control itself. When the program is required again, it can be easily retrieved from the floppy diskette.

For those CNC controls that do not have built-in floppy drive systems, there are a number of portable floppy-drive systems available from different manufacturers. All use the RS-232C (serial) communications port of the CNC control as described earlier. Figure 1.2 shows a portable floppy-disk drive unit.

The largest drawback of floppy diskettes in the machine shop environment is their susceptibility to magnetic fields. Many shops have

Figure 1.2 Portable floppy disk drive unit used to transfer programs to and from a CNC control. (*Courtesy GE Fanuc Automation.*)

strong magnetic and electromagnetic fields scattered throughout, and if the floppy diskette comes too close to them, the information on the diskette will be erased.

Floppy diskettes must also be kept relatively clean, free of the dust, dirt, and grime that accumulate around the typical CNC machine. If even one floppy diskette is damaged or inadvertently erased, it could mean the loss of hundreds of programs! For this reason, shops employing floppy diskettes for CNC program storage *must* make backup copies of their program storage diskettes on a regular basis.

Floppy-disk drive systems can be used to store and retrieve programs from any number of CNC machine tools as long as the machines have an RS-232C communications port. However, most companies do not use floppy-disk drive systems as their sole means of CNC program storage and retrieval (though some do). Most limit the purpose of their floppy-disk drive systems to simply transporting programs from a main computer to the CNC machine tools.

These portable program transfer devices are most often found in companies using a desktop computer in the office to prepare CNC programs. Maybe the distance from this computer to each of the machines is great, and the cost of running cables to each machine from the computer would be high. Since some floppy-drive systems use the same diskette formatting as PC-compatible computers, it is very easy for the programmer to transfer data between the floppy-disk drive system and the PC-compatible computer. Even for those floppy-disk drive systems whose diskette format is different from the PC, programs can still be transferred via the RS-232C (serial) port with relative ease.

The reason portable floppy-disk drive systems are not commonly used as a company's sole program storage and retrieval device is that most floppy-drive systems do not allow programs to be created or edited. In most cases, their sole purpose is to extract programs from the CNC machine and to send programs to the CNC machine. Most companies using CNC equipment need more than this when it comes to creating and editing their CNC programs.

Since a floppy-drive system has such a limited purpose, and since a laptop or notebook computer can be used for the same purpose for less than half the price, the portable floppy-drive system may not be the wisest choice for a program storage and retrieval system. Though the floppy-drive system may be better suited for the shop environment, since it is much more rugged than the typical laptop or notebook computer, the flexibility of a laptop or notebook is much preferred by most companies when a portable program transfer device is needed.

Portable program transport devices that use random-access memory

When transferring programs between a PC-compatible computer and a CNC machine, truly the most effective and efficient method is to run one cable to each CNC machine tool. However, as mentioned earlier, there are times when this is not altogether feasible. If the distance from the machine to the computer is over about 200 ft, the cost of the cabling may be prohibitive. And the more CNC machines a company owns, the greater this cost.

Also, if a company incorporates machine tools that cause a high degree of disruptive electrical surges in power lines (like electrical discharge machines or welders), cables run near these machines will be prone to interference. If this occurs, characters (and even program segments) could be lost during transmission.

In these cases, it may be necessary to manually transport the CNC program from the PC-compatible computer to the CNC machine tool. Though any portable program transfer device will work, the least expensive device used for this purpose is a random-access memory (RAM) CNC program-transporting device. Figure 1.3 shows a transport device that uses RAM memory.

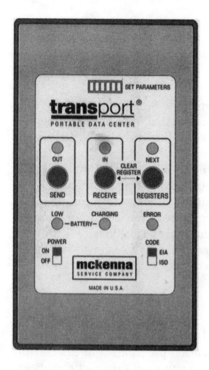

Figure 1.3 Portable random-access memory (RAM) device used to transfer programs to and from a CNC control. Courtesy (*Mckenna Service Company.*)

Most RAM-type program transfer devices are very small in size and are powered by a battery. Programs can only be retained by these devices while the power to the device is turned on. The programmer loads a program into the device from the PC-compatible computer. With the power kept on, the device is taken out to the CNC machine tool where the program is transferred into the control's memory. Once in the control's memory, the program can be cleared in the portable RAM device. Of course, programs can also be transported from the CNC control to the RAM transport device and eventually back to the PC-compatible computer.

Keep in mind that these devices can be used only to transfer programs, and that programs are *not* permanently stored. As soon as the device is turned off, the program is lost. Even so, they do make an excellent and inexpensive way for getting programs to and from the CNC machines.

Magnetic audiocassette tape recorders/players

Though becoming less and less popular, audiocassette tapes are still being equipped on some CNC machine tools as the primary method of program storage and retrieval. With this kind of device (either built into the CNC control or as a separate device), an audiocassette tape or microaudiocassette tape is used as the program storage medium. Normally, only two programs can be saved per cassette (one program per side of the cassette).

These devices tend to be cumbersome to work with and require quite a bit of time for program transfers. Also, like floppy diskettes, magnetic cassette tapes do not survive well in the dusty, dirty shop environment.

Since the stored program is extracted as a modulated audio signal, the volume (gain) setting is quite critical. Depending on the quality of the device, program loading can sometimes take several tries. This can make magnetic audiocassette program storage devices quite cumbersome and frustrating to work with. Unless this form of device is the only one the CNC control manufacturer allows, another, easier-to-work-with and more reliable form of program transfer device should be sought.

Laptop/notebook computers

Laptop and notebook computers make excellent program transfer devices, giving the user both flexibility and portability. Flexibility comes from the various software application programs that can be run. Today's small laptops and notebooks can run even the most complicated software applications. This means that CNC applications software such as text editors can be used to help with program preparation. Once a program is entered and stored on the hard drive or floppy drive of the laptop/notebook computer, it can be easily transferred to the CNC control via the serial port of the computer and the RS-232C port of the CNC machine tool. There are any number of software manufacturers that can supply such text editor and communications software (more on this later). With the help of a printer, CNC program listings can be generated from the text editor of the laptop or notebook computer. Figure 1.4 shows a picture of this type of computer.

Aside from simple text editors and communications programs, today's laptop and notebook computers can even run more sophisticated program preparation software packages. Even elaborate CAM (computer-aided manufacturing) systems can be run on many of these portable computers.

Since current laptops and notebooks can even run any software that runs on larger PC-compatible desktop computers, applications that are totally unrelated to CNC—like word processors, databases, and desktop publishing— are also quite applicable. When you consider that these small computers are sometimes less than half the price of a portable floppy-drive system that can *only* save and retrieve

Figure 1.4 A standard notebook or laptop computer can be used as a CNC program transfer device. (*Courtesy CNC Concepts, Inc.*)

CNC programs, you can see why more and more companies are using laptops and notebooks for program transfer instead of floppy-drive systems or other forms of program transfer devices.

The portability of any laptop or notebook computer rivals that of even the best floppy-drive systems. Most notebooks weigh less than 7 lb, and most laptops weigh less than 15 lb. However, keep in mind that laptop and notebook computers are much more fragile than the typical floppy-drive system designed specifically for CNC use, and as such, must be handled more carefully.

Desktop computers

The last form of program transfer device to be discussed is the desktop computer. These computers give every benefit of laptop and notebook computers except for portability. When desktop computers are used, most companies run a cable from each CNC machine they own to the computer. At the machine, the cable is connected to the RS-232C communications port. At the computer, the cable is connected to the serial port. Figure 1.5 shows a desktop computer.

Note that if more than one machine is to be connected to the desk-

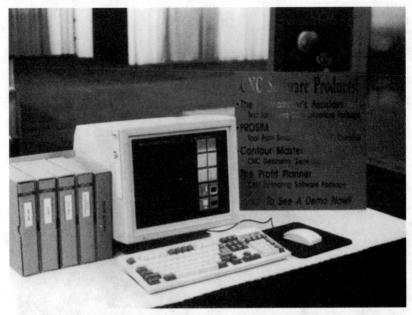

Figure 1.5 A desktop computer can be used as a CNC program transfer device. (*Courtesy Softwerks International.*)

top computer, some form of switch box must be incorporated. Many companies using this kind of system use a simple manual switch box (like an A-B or A-B-C switch) at the computer. In this case, the programmer simply selects the machine with which to communicate before the command to transfer a program is given. However, as the number of machine tools grows, manual switch boxes become somewhat cumbersome to work with. For more elaborate requirements, an automatic switch box can be used that will automatically detect program transfer commands. (More on this later, when we discuss distributed numerical control systems.)

As with the laptop or notebook computer, a variety of software application programs can be used, including CNC-related software as well as software unrelated to CNC. This gives the CNC user a great deal of flexibility when it comes to computer utilization.

Final note on program transfer devices

For the most part, we have discussed each device as it is applied when transferring programs directly to and from a CNC machine tool. But keep in mind that most of these devices can communicate CNC programs to *any* of the others. Most RS-232C–"speaking" devices can transfer programs to any other RS-232C–speaking device.

For example, say your shop has an older CNC machine that does not have an RS-232C interface. Say it can accept a program *only* in the form of a paper tape. A computer (notebook, laptop, or desktop) can still be used to prepare the program. However, an RS-232C–speaking (serial) reader/punch must be used to punch a tape of the program from the computer. This tape can then be used to load the program into the machine tool.

To get the program back from the CNC machine tool to the computer (after modification), a tape can be punched at the machine to a portable reader/punch. Keep in mind that if the machine does not have an RS-232C interface, a parallel tape punch must be used to extract the tape from the machine. This corrected program tape can then be loaded into the computer and stored on the hard drive or floppy drive via the (serial) reader/punch.

When considering the purchase of any new program preparation and transfer device (especially if it is replacing an older system), it is important for the company to maintain the ability to transfer programs to and from all of the various CNC (and older NC) machines the company owns. In most cases, this is easily possible if RS-232C–speaking devices are involved.

Devices for Creating and Editing Programs

You have now been exposed to the most common forms of program transfer devices. Of course, before any of these devices can be used, the CNC program must exist. In this section we intend to show the most common devices used to create CNC programs. Keep in mind that we are talking about *G code level* CNC programs at this time. It is from this level that the CNC machine will actually execute. Note that some program preparation devices allow the programmer to program at a higher level, incorporating graphics and simulations during the programming process. However, no matter how the program is prepared, the end result will be that a G code level CNC program is created.

During this section we will work from the simplest methods of creating CNC programs to the more sophisticated program preparation methods. Each company bases its decision as to which device is best for its particular situation on a number of different factors. As we go through this section, we will also give some basic criteria as to how companies make this important decision.

Entering and editing CNC programs at the control

All CNC controls allow the operator to enter and edit CNC programs using the keyboard and display screen of the control. This means the

CNC control itself can be used as the program preparation device. (Keep in mind that we are discussing G code level CNC programs and not addressing conversational controls at this time.)

Newer CNC controls allow programs to be entered as the machine is running its cycle and machining a workpiece. This feature is known as *background edit*. However, entering CNC programs from scratch at the CNC control is quite tedious, cumbersome, and error-prone, and most companies seek another method of entering programs.

If the company will never run the same job again (absolutely no repeat business), and if the operator is the person actually writing and entering the program, a case can be made for entering programs at the machine. Some job shop people, for example, will tell you that they *never* see the same job twice. Or they will say that, for the few times when repeat orders *do* come in, they simply build the program preparation and entering time into the cost of the job each time the job comes in.

In many job shops, the operator is the person who actually programs, enters, sets up, and runs the program. Since the operator can do only one thing at a time, some people feel that it does not matter whether the operator enters the program on line at the CNC control or through some off-line program preparation device. Since all program preparation devices do cost money, some job shop people feel they are actually saving money by entering programs at the CNC control. Figure 1.6 shows a picture of the display screen of a typical CNC control.

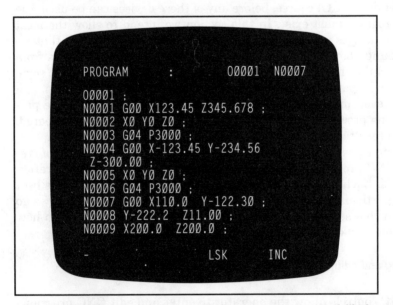

```
PROGRAM        :        O0001  N0007

O0001 ;
N0001 G00 X123.45 Z345.678 ;
N0002 X0 Y0 Z0 ;
N0003 G04 P3000 ;
N0004 G00 X-123.45 Y-234.56
  Z-300.00 ;
N0005 X0 Y0 Z0 ;
N0006 G04 P3000 ;
N0007 G00 X110.0  Y-122.30 ;
N0008 Y-222.2  Z11.00 ;
N0009 X200.0  Z200.0 ;

  -                     LSK       INC
```

Figure 1.6 Display screen of CNC control showing a program available for editing. (*Courtesy GE Fanuc Automation.*)

As you may be able to tell from the tone of this presentation, we tend *not* to agree with this line of thinking. There are several benefits from off-line program preparation devices that have nothing to do with repeat business. In general, these benefits reduce the time it takes to create CNC programs.

First, even very simple and inexpensive text editors that cost less than $1000 (to be discussed a little later) allow many editing functions that speed up the program-entering process. Automatic sequence numbering (for N words); cut, copy, and paste functions for repeated CNC commands; and even help with basic arithmetic and trigonometry calculations are available through most text editors oriented to CNC.

Second, the shop environment is usually dirty, noisy, and not a good environment for concentrating. An operator trying to enter a program at the machine is prone to be distracted and interrupted during the program-entering process. Conversely, the off-line program preparation device can be placed in a quiet room, away from the hustle and bustle of the shop. The operator will be allowed a much higher level of comfort and concentration.

Third, the placement of the control panel on most CNC controls is usually not a comfortable position to work with. Most control panels and display screens are mounted vertically, making the operator reach up to at least shoulder level. If it takes over about 10 minutes to enter the program, the operator's arm will fatigue. By comparison, most text editors utilize a keyboard similar to a typewriter, and the keyboard can be placed at table level. This makes it much more comfortable to enter the program.

Fourth, many CNC controls include only the letters of the alphabet needed to enter the CNC commands (like N, G, X, Y, Z, and so on). While this may sound just fine, there may be times when a message (in parentheses) must be included in a program. If all the letters of the alphabet are not available on the keyboard of the control panel (as well as the left and right parentheses), messages cannot be entered through the control. All off-line program preparation devices allow the typist to enter every character of the alphabet. Along the same lines, the placement of characters on some CNC control keyboards makes little sense. The placement of the characters on the keyboard of an off-line device will be like that of any standard typewriter.

As you can see, there are numerous reasons *not* to use the CNC control as a program preparation device. Think of it this way: A CNC machine makes a *very* expensive typewriter! Keep in mind that the CNC control manufacturer's reason for giving the user editing functions is to allow *existing* CNC programs to be easily modified. While an operator can also enter new programs, the CNC control makes a relatively poor choice as a program preparation device.

Program-entering functions of a CNC control. Generally speaking, most CNC controls allow only the most rudimentary program-entering and editing functions. While some do allow some of the complex editing functions found on more elaborate text editors, many CNC controls give only enough editing functions to allow efficient editing of existing CNC programs. These basic editing functions can be broken down into four basic categories: search, insert, alter, and delete.

Keep in mind that there will be variations from one control manufacturer to the next with regard to the specific keystrokes required to scan and edit within CNC programs. While we do give some rather specific discussions in this section, our intention is simply to acquaint you with the editing functions available. However, you must be prepared for variations.

Searching functions. All CNC controls give the operator a cursor with which to scan within the program. A cursor is a visible box or line that appears within the text of the CNC program and represents the current action point in the program. The first task required for almost any editing function is to move the cursor to the desired point. This is done with searching functions.

Most CNC controls allow this cursor to be moved within the program one word at a time with cursor forward and back, one line at a time with line forward and line back, and one whole page at a time with page up and page down. For relatively short programs, these very basic cursor controlling functions may be sufficient for scanning.

For longer programs, these elementary cursor control keys will be quite cumbersome to work with. For this reason, most CNC controls allow a more powerful way of scanning to the next single occurrence of any word (or letter address). Say for example, the programmer uses sequence numbers (N words) within the program. Say that the operator wishes to change something in line number N565 in the program. To search to this command with the page and cursor keys will be time-consuming and cumbersome. In this case, the operator can simply type the word N565 and press the key corresponding to forward search. The control will immediately search to the next occurrence of the word N565.

Most CNC controls also allow you to easily scan to the next occurrence of any word type. For example, say you wish to modify all feed rates in a program. Of course, the letter address for feed rate is F. The operator can simply type the character F and press the key corresponding to forward search. The control will immediately scan to the next occurrence of an F word.

Insert. This function is used heavily when programs are entered into the control from scratch at the machine. As the name implies, this function allows new CNC words and commands to be registered into the program. If typing in a new program from scratch, the operator will enter each command followed by an end-of-block (EOB) character and then press the key corresponding to insert. If a mistake is made during the typing of this new word or command, the operator will have a backspace or cancel key with which to back up one character in the entry. Most controls will place the inserted information *after* the current location of the cursor. If an entirely new program is being typed, the current position of the cursor will be the last word of each command typed. So the operator can easily continue entering program words and commands without concern for the cursor. If adding new words or commands to a current program, on the other hand, the operator *must* first position the cursor to the last word *prior* to the intended location of the new program information.

Alter. This editing function allows words in the program to be changed. Possibly the operator meant to enter a word as X4.5 but made a mistake and entered the word as X5.4. In this case, the program is incorrect and must be changed. To fix this problem, the operator *must* first position the cursor to the incorrect word (X5.4 in this case). With the cursor on this incorrect word, the operator simply types the new (corrected) word and presses the key corresponding to alter.

Delete. The last of the rudimentary program entering and editing functions is delete. This function allows the operator to remove a single word, a single command, or a group of commands from the program. In fact, entire programs can be deleted from the CNC control's memory. Keep in mind that, once deleted, the information is gone for good. If the operator makes a mistake during a delete function, program information can be inadvertently lost. For this reason, the delete function tends to be the most dangerous of the program-editing functions, and the operator must be very careful with its use.

To delete a single word in memory, the operator simply scans to the word to be deleted (brings the cursor to the word) and presses the key corresponding to delete. Some controls will ask for a conformation at this point to assure that the operator knows what is going to happen.

To delete an entire command, the operator scans to the first word of the command to be deleted (usually an N word) and then types the end of block (EOB) key and presses the key corresponding to delete.

To delete a series of commands, the operator scans to the first word of the first command to be deleted. He or she then types the sequence number (N word) of the last command to be deleted. When the key corresponding to delete is pressed, the entire series of commands is removed from the program.

Final note about editing functions. If you are using the CNC control as your program preparation device, keep in mind that you are working on your *only* copy of the program as you enter and edit your program. Most other forms of program preparation devices let you easily create backup copies of your programs. If you make a drastic mistake when editing a program, the backup of your program will remain intact.

Unfortunately, most CNC controls do not allow you to easily make backups without the help of some other device. Note that any one of the program storage and retrieval devices discussed earlier will allow you to make a backup.

Our point is this: if you are working on the only copy of the program, you must be extremely careful when editing, so as not to damage the program. If dramatic mistakes are made during editing, much time can be wasted while trying to get the program back to its original state!

Understanding background edit. As stated, a CNC machine tool makes a very expensive typewriter. If it is possible, it is wise to find a way to create your CNC programs off line. This will minimize the CNC machine's downtime while programs are loaded. When it is necessary to enter programs at the machine through the keyboard of the CNC control (as is the case in some job shops), there is a feature, mentioned earlier, called *background edit*. This feature helps to minimize downtime while the CNC program is being typed into the control.

Background edit (sometimes called *simultaneous edit function*), allows the operator to be typing a CNC program as the machine tool is running production. An operator can load the workpiece and execute the machining cycle. Once the cycle is running, the operator can select the background edit mode and begin (or continue) entering a new program.

One limitation of this feature has to do with cycle time. If a workpiece with a lengthy cycle is being run, the operator may have ample time in which to enter a new program. Long-running machining-center programs and bar feed applications on turning centers are two such instances of when the operator will have a great deal of time during which to enter the next program.

If, on the other hand, the machining cycle is quite short (say only a few minutes), it is likely that only a small portion of the next program will be entered at the time when the next workpiece must be loaded. In this case, the operator will be constantly breaking out of the program-entering procedure to load and measure workpieces as well as to make the necessary offset adjustments required from workpiece to workpiece. This can be very frustrating to the operator, and opens the door to typing errors, since the operator's train of thought is continually broken.

What about program storage? When using the CNC control as a program preparation device as shown above, keep in mind that there still may be the need to store and retrieve programs. That is, the CNC control can hold only a limited number of CNC programs before it becomes full. If the company truly knows that a workpiece will never be run again, it might elect to delete the program for a workpiece without concern. Of course, this means that all program information about the job will be lost, and if the job *does* happen to come into the shop again (or a very similar job), it will mean a duplication of effort. For simple jobs, this may not represent a great deal of wasted time, but for more complicated jobs, the time wasted can be substantial.

Keep in mind that all of the program storage and retrieval devices discussed earlier in this chapter can be used to store CNC programs, even if they are created at the CNC control. If cost is the main issue, as would probably be the case if the CNC control is the program preparation device, we would recommend either a floppy-disk drive system or a laptop/notebook computer. Either would allow programs to be stored and retrieved, but in our opinion, the computer would make the wiser choice for two reasons. First, these small computers cost less than the typical CNC portable floppy-drive system. Second, they allow more flexibility, allowing the program to be entered or edited off line with a text-editor software program.

CNC text editor

This program preparation device has numerous advantages over entering CNC G code level programs at the machine. The most basic advantage of this kind of system is that programs can be entered off line. While some CNC controls offer this ability if they have the background edit function, generally speaking, the machine shop does not make a very comfortable or convenient environment in which to type CNC programs (for all of the reasons given in the last section).

Once a program is entered at the text editor, the actual time it takes to transfer the CNC program to the CNC machine tool (through the RS-232C port) is little more than a few seconds. This means the actual CNC machine downtime due to program loading can be kept to a minimum.

We will discuss the various features and utilization of RS-232C communications a little later, during our presentation of distributed numerical control. For now, we will address *only* the CNC text editor itself. Figure 1.7 shows the display screen of a common CNC text editor.

Most current CNC text editors are personal-computer–based. Even the lowest-level personal computers available today offer more than enough power for even the most elaborate text editors. This is because CNC text editors are "text-based" and require little or no graphics or "number crunching" as may be required by other computer applications like databases and computer-aided design (CAD) systems. For this reason, the system requirements for a PC-based text editor are minimal, and even the most inexpensive personal computer (PC) will usually suffice as a CNC text editor.

If you have experience with personal computers, you can think of a CNC text editor as being like a word processor oriented to the CNC environment. Most CNC text editors incorporate many of the same global editing features found in word processors like find and replace, cut and paste, and copy and paste. In fact, any word processor can double as a CNC text editor *if* it has the ability to save and recall programs in the text file format called *ASCII format*. Most popular word processors sold today have this capability.

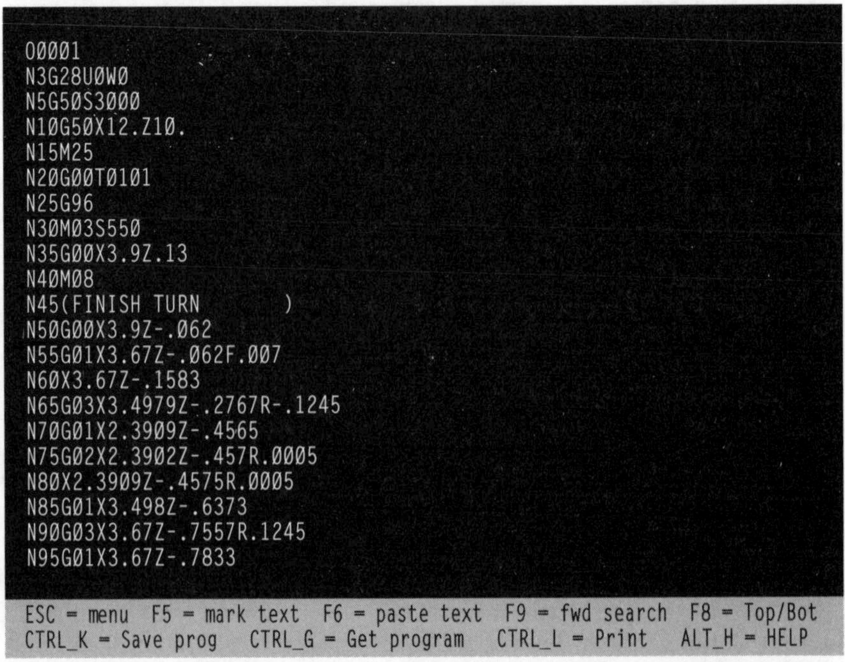

```
O0001
N3G28U0W0
N5G50S3000
N10G50X12.Z10.
N15M25
N20G00T0101
N25G96
N30M03S550
N35G00X3.9Z.13
N40M08
N45(FINISH TURN        )
N50G00X3.9Z-.062
N55G01X3.67Z-.062F.007
N60X3.67Z-.1583
N65G03X3.4979Z-.2767R-.1245
N70G01X2.3909Z-.4565
N75G02X2.3902Z-.457R.0005
N80X2.3909Z-.4575R.0005
N85G01X3.498Z-.6373
N90G03X3.67Z-.7557R.1245
N95G01X3.67Z-.7833

ESC = menu   F5 = mark text   F6 = paste text   F9 = fwd search   F8 = Top/Bot
CTRL_K = Save prog   CTRL_G = Get program   CTRL_L = Print   ALT_H = HELP
```

Figure 1.7   Display screen of a CNC text editor. (*Courtesy Softwerks International.*)

Keep in mind, however, that creating and storing your CNC programs is only *half* of what the CNC text editor is intended to do. You must also have the ability to transfer your programs between the computer and the CNC machine tool. This means that a common word processor will perform only half of what a true CNC text editor will do. Another software package would be required to handle communications. We will show a very inexpensive way of communicating programs created with any ASCII word processor a little later.

CNC text editor features.   Since a CNC text editor is like a word processor oriented to CNC, we will discuss its editing features in two categories, word-processor-oriented features and CNC-oriented features. After the editing features, we will also address the communications features of CNC text editors.

Keep in mind as we go through these features that software suppliers vary as to how these features are applied. Our intention will be limited to introducing the feature and discussing how it applies to the task of entering and editing CNC programs.

Word-processor-related features.   If you have experience with word processors as they are used on personal computers, you will find many of these features to be familiar and easy to understand. If you have no previous computer experience, this presentation should go a long way to acquaint you with what is possible.

Entering new programs from scratch.   All CNC text editors allow the programmer to enter new programs at the text editor instead of through the CNC control's keyboard. Indeed, this is one of the most basic features of the CNC text editor in the first place. When you first activate the software, most CNC text editors will assume you wish to enter a new program as the first order of business. Generally speaking, if you do wish to enter a new program, you simply begin typing the program as soon as the software is running.

Scanning, insert, alter, and delete.   With any CNC text editor, you have the same rudimentary editing functions found within a CNC control. During or after the typing of a program, you have the ability to easily monitor and manipulate the commands typed. As with a CNC control, you will have a cursor that can be moved throughout the program. You will have cursor- moving keys that allow you to easily move the cursor forward and back one character, one line, or one page. If you wish to add characters in the program, an insert key controls whether your new typed characters are inserted into the program or whether they replace current characters (alter). A delete key allows you to remove characters from your program.

One minor advantage a CNC text editor has over editing at the

CNC control is that the text editor allows single characters of your program to be manipulated. Most CNC controls allow only changes to words, meaning several characters of your program must be typed in order to make a change. For example, say the word X15.376 is currently in your program, and you want to change this word to X15.375. With a text editor, you will simply place the cursor on the 6 of X15.376, be sure the insert key is set to replace (instead of insert), and type the replacing character 5. On the other hand, most CNC controls would require that you type the entire X word, X13.375, in order to make the change.

Saving programs. Once a program is entered, CNC text editors give you a way to store your program for future use. You can store your programs on floppy diskettes or, if one exists, on the computer's hard drive.

It is advisable to save lengthy programs on a regular basis during typing. Since most computers lose power during a power outage, any information typed since the last save will be lost if the power goes out. If you are close to the end of a 1-hour program-entering session when the power goes out, everything you typed will be lost. We recommend resaving your program every 10 minutes or so. Most CNC text editors have a way of easily commanding the resaving of programs.

Recalling programs for editing. It is very easy to recall a previously entered program for modification. During the recalling of a program, most current CNC text editors will show you a list (directory) of the programs currently stored on your disk drive. To recall a program, you will simply bring the cursor to the program to be recalled and press the enter key.

Easy backup. As mentioned earlier, it is advisable to make backup copies of all important CNC programs. Most CNC text editors allow you to easily designate the device to which a program will be saved, making the backup procedure as easy as commanding the save function twice. For example, say you wish to save all of your programs to your computer's hard drive. But you also wish to save your important programs to a floppy diskette as a backup. During the saving function, you simply designate to which device you wish to save and save the program twice, once to the hard drive, and once to the floppy diskette.

Making one program from another. Almost all companies have at least one series of workpieces that could be considered a family of parts. As you can imagine, CNC programs for the various workpieces in the family are very similar. In many cases, a program for one workpiece can be slightly modified to create a program for another workpiece in the family. CNC text editors make changing one program to create another an easy three-step procedure.

First, a program for one workpiece in the family is created and saved. Second, this program is modified to fit the criteria for a similar workpiece. Third, the modified program is saved with a new name. At this point two programs will exist, one for the original workpiece in the family, and one modified program for another workpiece in the family.

Find and replace. All CNC text editors allow several "global," or mass-editing, functions. Find and replace is one such feature. With this function, the programmer can command that all occurrences of a certain word be modified to become another word. For example, say a program runs a tool at a feed rate of 5.0 in/min. This feed rate (F5.0) may be commanded numerous times throughout the tool. Say the programmer wishes to increase all feed rates of F5.0 for this tool to F6.0 (6 in/min). One way to do this would be to cautiously move the cursor through the program and manually change each feed rate. This technique would be tedious and error-prone.

The feature *find and replace* will allow the programmer to easily specify that all occurrences of F5.0 be changed to F6.0 in one easy command. Most CNC text editors will even display each potential change for confirmation (if requested) to assure that each change is truly desired. Any CNC words can be easily changed in this manner.

Copy and paste. This mass-editing feature allows the programmer to easily repeat a series of CNC commands any number of times within the CNC program. The copied commands are placed into a temporary buffer. The information in this buffer can be easily placed (pasted) into the program at any cursor location.

This feature even allows a series of commands used in one program to be copied into another program. The programmer will simply load the program containing the commands to be copied and copy the commands into the buffer. Then the program into which the commands must be copied is loaded. The cursor is brought to the desired position and the copied commands are "pasted" into the new program.

Though the actual techniques needed to do this vary from one CNC text editor to the next, most text editors require that the commands to be copied be "marked" in one fashion or another. Once marked, the CNC text editor will give you the option of copying the commands or "cutting" them from the program. Either way, the commands are sent to the copy buffer for use later (when they can be pasted at some other position in the program).

Cut, and cut and paste. Like copy and paste, cut and paste utilizes a temporary buffer. But instead of simply copying commands into the buffer, cut by itself actually removes the series of commands from one section of the program. Note that if a whole series of commands is simply to be deleted from the program, the cut command can be used (no pasting later).

This feature is also used to move commands from one area of the program to another. This is most helpful when you have a problem with a program's sequence of operations (process). If you find that you must change the machining order, you can cut the programmed commands for a tool from one section of the program and paste it into another section of the program. This makes it very easy to change a CNC program's machining order.

CNC-related text editor features. As stated earlier, a CNC text editor is a word processor oriented to CNC. Now that we have discussed the word processor features of CNC text editors, here are the features of CNC text editors that are directly related to CNC. Figure 1.8 shows the main menu of CNC functions possible with one popular CNC text editor.

Automatic sequence numbering. To avoid the laborious task of typing a sequence number (N word) in every command of your program, CNC text editors allow you to completely ignore your sequence numbers as the program is typed. Once the program is completed, one simple command easily and automatically adds sequence numbers to the program. With this feature, the programmer even has the ability to resequence the numbering pattern and have each sequence

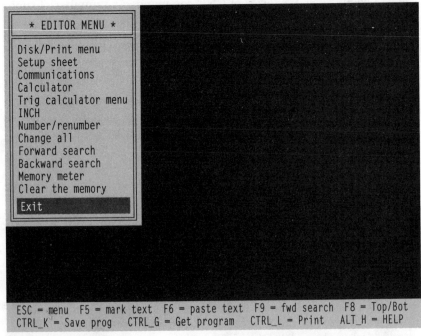

Figure 1.8 Main menu of a CNC text editor. Note the CNC related functions. (*Courtesy Softwerks International.*)

number skip a certain number between lines. For example, if the skip increment is set to five, the sequence numbers will come out as N005, N010, N015, N020, and so on. Many programmers prefer this sequence number structure to allow extra commands to be added within the program (at some later date) and still include sequence numbers in the added commands.

Arithmetic calculations. Most CNC text editors give the programmer some help for calculating the complicated coordinates required for the program. Though they vary with regard to how much help they give, most will include at least basic arithmetic calculations like add, subtract, multiply, and divide. This minimizes (or may even eliminate) the need for an electronic calculator as the program is prepared.

Some CNC text editors also include trigonometry functions. For example, the programmer simply types in the two known values of a right triangle and the CNC text editor responds by giving all other data about the triangle.

There are even some CNC text editors that allow the programmer to easily come up with the coordinates for complex contours. The programmer defines the geometry of the contour to be machined, and the CNC text editor responds by giving all pertinent data about the contour. Note that this borders on what a computer-aided manufacturing system is designed to do. More on CAM systems a little later.

Menu-driven CNC cycles. Many CNC text editors give the programmer a way to automatically generate the CNC commands for certain commonly used machining operations. For those operations the CNC text editor is designed to handle, creating the CNC commands is as easy as filling in the blanks for the operation.

For example, say the CNC text editor includes a bolt-hole circle menu for generating the X and Y coordinates for all holes on the bolt-hole pattern. The CNC text editor will ask for information such as the center coordinates of the bolt-hole pattern in X and Y, the number of holes, the radius of the bolt-hole circle, and the starting angle. From this limited information, the control will be able to calculate all coordinates needed in the program. Once finished, the CNC text editor will automatically place these coordinate positions into the program.

Other examples of special cycles given by CNC text editors include pocket milling cycles (round, square, and oval), other hole pattern commands (window pattern, line pattern, grid pattern), and grooving. Keep in mind that CNC text editors vary dramatically with regard to how many of these special cycle menus they give as well as how the information about each cycle is entered.

Speed and feed calculations. Many CNC text editors allow the programmer to easily calculate the speed in RPM and the feed rate in inches per minute (IPM). For RPM, the programmer will enter the desired surface feet per minute (SFM) speed as well as the machining diameter. The CNC text editor will respond with the appropriate RPM. For feed rate in IPM, the programmer enters the previously calculated RPM as well as the desired inches per revolution (IPR) feed rate. The CNC text editor will respond with the appropriate feed rate in inches per minute. Once calculated, the speed in RPM and the feed rate in IPM will be automatically inserted into the CNC program. These techniques are especially helpful on machining centers, since most machining centers can only accept speed in RPM and feed rate in IPM.

Redundant format insertion. Much of a CNC program is nothing more than formatting. All programs for the same CNC machine start in the same basic manner, meaning the program start-up format will be basically the same for every program written for the CNC machine. In the same way, tool changing format is basically the same from tool to tool, as is the basic format needed to end each program. As you may know, these various formats are very redundant, and must be repeated many times for the numerous programs a programmer writes.

To help with these redundant commands, many CNC text editors allow the programmer to store a series of commands that makes up each type of format. Normally, at least program start-up, tool changing, and program-ending formats can be stored. To command that these redundant commands be inserted into the program, only one or two key strokes are required. This feature also helps with other redundant commands such as bar feed commands on turning centers and pallet changing commands on machining centers.

Some CNC text editors call the ability to store these commands *keyboard macro* techniques. Others call them format *stamps.* No matter what they are called, they can dramatically reduce the number of commands a programmer must type.

Setup sheet generation. Many CNC text editors allow the programmer a way of generating setup sheets. Though most do not allow setup drawings to be made, they do allow the user to easily specify the tools (and tool stations), offsets, cutting tool inserts, and other information related to the program's tooling. Most also allow a written set of setup instructions to be entered and included on the setup sheet. Of course, this setup information can be saved on the hard drive or floppy diskette and printed to a line printer.

Easy program conversion. Some CNC text editors allow programs to be converted from a format required on one CNC machine to that required for another. Features related to this function include the ability to switch from decimal point programming to fixed format (and vice versa) and switching from radius programming (the R word) for circular movements to directional vectors (I, J, and K).

Communications. Once a program is created, CNC text editors give the user a way of transferring the program to and from the CNC control. This is done by using the serial port of the computer and the RS-232C communications port of the CNC machine tool. How the transmissions are made and the protocol for RS-232C serial communications will be discussed later, during the presentation of distributed numerical control systems.

A *very* inexpensive CNC text editor and communications system. A typical CNC text editor and communications software package written specifically for use with CNC will cost from about $150 to over $1000, depending on the supplier and the features of the text editor. Keep in mind that these prices are for software only, and do not include the computer or cable needed to complete the system. If cost is of primary concern, keep in mind that there is an inexpensive way to avoid having to buy a text editor specifically developed for use with CNC. Here is a way to create your own CNC text editor and communications system that takes advantage of very inexpensive generic software.

Hardware requirements. If you have *any* PC-compatible computer (including laptops, desktops, and notebooks), it is more than likely that you have all the hardware needed to create this system. Most computers sold today come standard with at least 640K internal memory, floppy drives and/or hard drive, keyboard, serial port, and monitor. Most people buying computers will also buy a printer. If your existing computer has these features, you can create a simple and inexpensive CNC text editor and communications system. Keep in mind that this system will only have the word-processor-related features of the CNC text editor and limited communications ability. Even so, you will be able to easily create, edit, transfer, and store your CNC programs with this system.

Software requirements. You will need two distinctly different software packages. (Note that both of these software packages are included within any CNC text editor.) First, you will need a word processor or text editor that has the ability to save programs onto the hard drive or floppy diskette in ASCII format. Most current word processors have this ability, so if you are currently using a word processor with your computer, it is more than likely that you have this ability (check your word processor manual to find out).

If you do not currently own a word processor, keep in mind that most versions of DOS (your disk operating system) include a text editor that allows you to create and save documents in ASCII format. This text editor will be more than adequate, and we would recommend using this DOS text editor rather than purchasing a separate word processor if cost is of primary concern. Check with your DOS manual for more information about how to use this text editor. Though word processors do vary in price, they range from under $100 to over $300. By comparison, DOS comes standard with most computers, meaning you should not have to pay extra to get the text editor included with DOS.

The second software package you will need is a communications software package. Though the main purpose for these generic communications software packages is to transfer files from one computer to another, most can be adapted to transfer programs between the computer and any CNC machine that has an RS-232C serial communications port.

When we discuss distributed numerical control systems later, we will give the RS-232C related protocol requirements of your communications software. With this information, you can easily ask your software supplier to recommend a specific communications software package that will work. The current price for this communications software is under $75, and it is available from most computer stores.

Other considerations for your homemade CNC text editor system.   You will need a cable to connect the computer to the machine. The parts needed to make a cable up to 50 ft long can be purchased at any electronics store for $40 or less. When we discuss distributed numerical control systems a little later, we will show the actual cable diagrams.

If making this kind of system for yourself, be prepared for some legwork. *You* will have to make the cable, learn the text editor software, learn the communications software, and tailor the RS-232C communications protocol for both the CNC machine and the communications software by yourself. If you have problems, you'll be on your own to figure them out. Though we intend to show you enough during our discussions of distributed numerical control for you to be able to do this, it is *not* a task that can be done in a few minutes. But if you're willing to take on the challenge, and if you currently have the hardware (computer), you can make your CNC text editor and communications system with one cable for an additional cost of under $150!

If cost is not of primary concern, if you value your time, and/or if you do not wish to be bothered by doing the related legwork, we encourage you to purchase one of the many excellent software packages written specifically for CNC application. Though more expensive, they will make it *much* easier to get the system up and working. Most CNC text editor suppliers are willing to supply you with the cables and tell you exactly how to tailor the RS-232C communications protocol for your particular CNC controls. To find suppliers for CNC text editors, you need only pick up a trade publication oriented to the machine shop environment. These journals are filled with ads from such suppliers.

Compatibility with other program storage and retrieval devices. As with any serial (RS-232C) device, a CNC text editor can be used to transfer programs to any other RS-232C–speaking device. There are two situations when this knowledge is especially important.

One time this is necessary is for older CNC (or NC) machines that can only accept programs from paper tape. In this case a (serial) tape reader/punch can be used to punch tapes directly from the CNC text editor. The program (on tape) can then be brought to the machine and run. To retrieve programs from the machine, a tape must be punched from the machine tool to a tape punch. This program (on tape) can be brought back to the computer and loaded back into the CNC text editor from the reader/punch.

Another time it is helpful (if not mandatory) to transfer programs to and from the CNC text editor to another RS-232C device is when it is a great distance from the computer to the CNC machines. If the distance is great, some companies elect to hand-carry programs temporarily stored in another RS-232C device from the computer to the CNC machine. Portable computers (laptops and notebooks), portable floppy drives, and random-access memory (RAM) transport devices are commonly used for this purpose.

Conversational CNC controls

The third level of program preparation device is the conversational CNC control. This type of control allows the CNC machine operator to input the program right at the machine tool. But instead of entering a G code level or manual program, the operator is allowed to enter a more interactive program. Though conversational CNC controls vary dramatically with regard to their level of sophistication and as to how they are actually programmed, most give many features that effectively reduce the amount of work an operator must do in order to prepare a CNC program. Figure 1.9 shows the display screen of one popular conversational CNC turning center control.

Before we dig in too deep, we must point out that there has been quite a controversy brewing over the logic behind applying conversational controls. Many people question the wisdom of shop-floor-programming a CNC machine.

As with any controversy, there are certain people who are very much in favor of using conversational controls and those who oppose their use. Though there are exceptions to this statement, generally speaking, smaller companies (job shops and small manufacturing companies) tend to be more in favor of utilizing conversational controls than larger manufacturing companies. Let's look at the reasons why.

As mentioned earlier in this chapter, smaller companies tend to

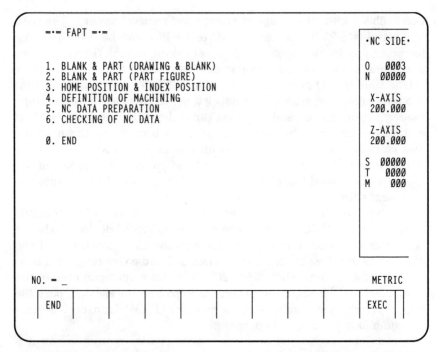

Figure 1.9 Display screen of a common conversational turning center control. (*Courtesy GE Fanuc Automation.*)

utilize their personnel in a different manner than larger companies. In the smaller company, a CNC operator is required to do more than simply run production. In small companies, the CNC operator may be expected to develop the process plan for a job, write the program, enter the program, make the setup, verify the program, *and* run production. Once production is being run, the operator may even be expected to inspect each workpiece and make machine adjustments accordingly. In this case, the operator wears many hats, and acts as process engineer, tooling engineer, programmer, setup person, CNC operator, and inspector.

If the operator is utilized in a great number of ways, anything that can be done to reduce the workload will be desirable. From the time one job is finished to the time the next job is totally verified and running production, the CNC machine tool will be sitting idle, not running production, and of course, *not* making money for the company.

If conventional (manual) programming techniques are used, the programming process can be one of the more time-consuming tasks the CNC operator must perform. For complex workpieces, and if production quantities are low, it could easily take longer to program a workpiece than to set up and run production! Conversational controls take much of the tedious and error-prone work out of CNC programming.

On the other hand, larger manufacturing companies adhere to a dramatically different philosophy. They believe that any machine tool (CNC or otherwise) should be running production at all times. These companies view *any* machine downtime as lost machining time and take great pains to assure that the machine tool runs workpieces for as great a percentage of time as possible. They hire a CNC support staff that is adequate to maximize production time. In this kind of company, several people are involved in the CNC environment, and the CNC operator's responsibility is strictly limited to running production.

For any one job to be run on the CNC machine, a process engineer prepares the machining process. A cutting tool engineer checks, designs, and/or orders the cutting tools. A tool designer designs and supervises the construction of the work-holding tools. A CNC programmer prepares the program. A setup person makes the work-holding setup, loads the CNC program, and runs the first workpiece. An inspector checks the first workpiece and gives the go-ahead to run production. Finally, the job is turned over to the CNC operator, whose responsibility is limited to loading workpieces and executing the cycle.

This situation, of course, is much better geared to higher production. Setup (nonmachining) time is dramatically reduced by the support people. As production is being run, future jobs are constantly being prepared. Compare this to the small company example given earlier where the operator may not even *begin* to get ready for the next job until the current job is finished.

As you can imagine, larger manufacturing companies tend to see shop floor programming on conversational controls for their production machines as a waste of time. This kind of company may utilize conversational controls in their tool room, for tool-making and prototype purposes, but never in their production environment.

Features of conversational controls. For those companies that need their CNC machine operators to prepare CNC programs, conversational controls make an excellent alternative to manual programming. As we present the features of conversational controls, you should begin to see why.

CNC side versus conversational side. Most conversational controls have two computers, one that acts as the CNC control and the other that acts as the program preparation system. This kind of conversational control gives the user the best of both worlds. If the user desires, this style of conversational control could be used as a standard CNC control (as a larger manufacturing company would like). That is, CNC programs prepared in *any* manner could be loaded into the control and run as usual. The conversational system does *not* have to be used to prepare the program.

Conversely, since conversational controls make it much easier for the *operator* to prepare programs, these controls facilitate shop floor programming (as preferred by smaller companies). In this case, the CNC program will be generated by the conversational control.

Note that since most conversational controls truly have two different computers, most allow the operator to be inputting a conversational program on the conversational side *while* the machine is running a workpiece on the CNC side. Of course, this would only be feasible during longer- running machining cycles.

For this reason, conversational controls tend to be a more versatile choice for any machine purchase, even for machines that will be used in a production environment. If, at some later date, the CNC machine tool is to be used in another area of the company (maybe the toolroom), they will facilitate shop floor programming. And, generally speaking, they will tend to give the machine tool a higher resale value when it comes time to replace equipment.

Graphics. Most conversational controls give the operator a very good view of what is happening during the entire programming process. They let the operator visually check all important functions of the program being generated. There are two areas when this graphic capability is most helpful.

First is during the definition of the workpiece. At some point during the conversational program, the operator will describe the finished workpiece. During this description, most conversational controls will actually draw the workpiece on the display screen. The operator will be able to see just what the control thinks the workpiece looks like.

The second time this graphic display is especially helpful is during (or after) the description of machining operations. Most conversational controls will display a tool path, showing the operator exactly what each tool is going to do. Some even animate the machining operations. These sophisticated conversational controls will actually show each tool removing material from the workpiece.

When the operator is finished with the program, there will be no doubt as to what each tool is going to do. The operator will have seen the finished workpiece drawing on the screen and seen how each tool will do its machining. While mistakes may still have been made, the potential for major problems with the program is greatly reduced. Compare this to the poor manual programmer who is working at G code level, where a CNC program is nothing more than a series of cryptic CNC commands. Unless some form of program verification device that can plot CNC programs is used, the programmer will have no idea as to what the program will really do until it is run on the machine tool.

Advanced error checking. As you can imagine, the graphic capability of the control minimizes the potential for making mistakes. If a mistake is made during the drawing of the workpiece, the drawing on the display will not look right. If a mistake has been made with a depth of cut during machining, the operator will be able to see this during the tool path simulation. Many such visual checks can easily be made.

Aside from this visual checking, most conversational controls have a great deal of "intelligence" regarding the functions the machine tool is intended to perform. If the operator breaks a rule related to the machine's function, most conversational controls will respond in a way that lets the operator know something is wrong. Possibly an alarm will be generated, or possibly the control will not react in the expected way.

For example, say an operator is working with a turning-center conversational control. Say this operator is trying to turn the outside diameter of a workpiece. Say he or she is finished drawing the workpiece and intends to begin the definition of machining operations. As the operator selects the tool to do the rough turning, say he or she makes a mistake. Instead of selecting a turning-style tool, a boring bar is selected instead. Of course, a boring bar is not capable of turning the outside diameter of the workpiece. Most conversational controls would *not* let the boring bar machine on the outside diameter of a workpiece. On most, an alarm would be generated.

Keep in mind that this example is just scratching the surface of the kind of error checking most conversational controls include. Hundreds and hundreds of machine, tooling, and cutting condition functions are checked in this manner by most conversational controls. While there are still loopholes, and programming is not yet fail-safe, the potential for making mistakes is dramatically reduced.

Reduced math. Most CNC programmers would agree that manual programming involves a great deal of math. Every motion command of the program requires at least one axis departure. And every axis departure requires an end point for the motion. Because most conventional (not conversational) CNC controls require coordinates exclusively in the rectangular coordinate system, every end point (in at least one axis) must be calculated for every motion command. In some cases, calculating end points simply involves adding and sub-

tracting dimensions on the workpiece drawing. But the more complicated the series of motions, the more complicated these calculations.
In many cases, a knowledge of trigonometry is required for making
end point calculations manually.

Conversational controls greatly reduce the need for this math
while programming. In fact, most conversational control operators
will not have to do *any* math while programming. Even for extremely
complicated contours, describing the contour to the conversational
control will be as simple as entering values taken directly from the
workpiece drawing. In other words, the operator is not required to
work exclusively in the rectangular coordinate system while programming as is the case in manual programming. Instead, the operator simply describes the attributes of the workpiece using a more
graphic series of techniques. Points, lines, and circles that represent
surfaces of the workpiece are drawn using logical and easy dimensioning techniques. For example, angular surfaces are represented by
lines and allow the angle of the line to be input as part of the line definition. When actually generating the CNC program, the conversational control will automatically calculate the rectangular coordinates required for each motion within the CNC program.

Reduced requirements for cutting conditions. Almost all conversational
controls incorporate some form of material database (usually called
the material file) that causes the control to automatically come up
with the cutting conditions required for each machining operation.
The end user can enter the materials machined by the company
when the CNC machine is new. From then on, whenever running one
of the materials included in the material file, the conversational control will come up with the proper feed rates, spindle speeds, and
depths of cut as the program is input. This dramatically reduces the
time it takes the operator to look up the cutting conditions in a cutting conditions reference manual.

Keep in mind that the cutting conditions generated by the material
file during programming are nothing more than a starting point. The
operator will have total control of the speeds, feeds, and depths of cut
as the program is written. An operator who does not agree with these
automatically generated machining conditions can easily change
them while entering a program.

Tooling considerations. In like manner, most conversational controls
allow the user to organize tooling in a tooling database (usually
called the *tool file*). This allows the control to memorize information
about the tooling used by the company. This tooling information is an
essential part of the conversational control's ability to machine with
each tool correctly. Usually, the tooling information describes the

geometry of each tool so the control can determine what the tool is capable of machining.

For example, most turning-center conversational controls require that (at least) the nose radius, lead angle, and nose angle be described for all single-point turning tools. This geometry tells the control (among many other things) to what degree recesses can be machined. In many cases, this tooling information knowledge keeps the control from doing something that is not possible because of tooling limitations.

Reduced need for machining practice knowledge. Most conversational controls take away much of the need for the operator to possess a high level of machining practice experience. Many machining operations are built in; the operator can simply turn loose the control to perform the operation, and the control will do so correctly. This is *not* to say the operator of a conversational control does not need a basic understanding of machining practice. The more an operator knows about machining practice, the better. Our point is that the conversational control will reduce the amount of thinking an operator has to do with regard to machining practice in order to prepare a program.

For example, most machining-center conversational controls have a series of pocket-milling commands. Once the pocket is described (length, width, depth, and so on), the operator simply tells the control the diameter of the end mill to be used to machine the pocket. This diameter, along with some very basic cutting conditions, like depth of cut and percentage of overlap, give the conversational control all it needs to completely machine the pocket. The operator does not have to plan every pass the end mill will make. The conversational control will do so automatically.

General flow of conversational programming procedure. Conversational controls vary when it comes to how programs are input, depending on the application for the control (machining center, turning center, etc.) and the control manufacturer. Our intention in this section is to present the three most common steps in the typical order of input required by most conversational controls.

General information. The first step to conversational programming is giving the control some basic information about the program to be input. Much of this general information sets the criteria for how the workpiece should be drawn and machined later. Figure 1.10 shows the general information required by one popular conversational control.

Here is a description of the most important general information requirements of conversational controls.

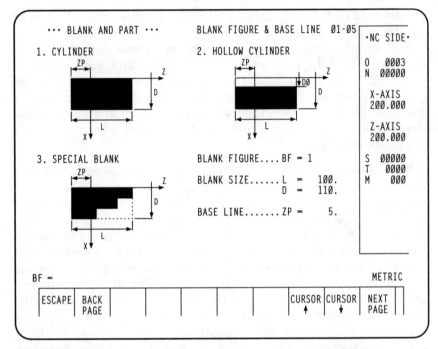

Figure 1.10 Display screen of a common conversational control showing the general information required during the first step of conversational programming. (*Courtesy GE Fanuc Automation.*)

Workpiece material. In order for the control to determine cutting conditions (speeds, feed rates, depths of cut) for each machining operation, the control must know the material to be machined. Most conversational controls have a material database (called the material file) that includes standard criteria for each cutting operation. Keep in mind that these standard values are just beginning points. The operator can easily change any cutting condition value as the program is written.

Say for example, you are working with a turning-center conversational control. In the material file, you enter the material "low carbon steel." In this material, you will enter speeds, feed rates, and depths of cut for each machining operation (like rough turn, rough bore, finish turn, finish bore, and threading). Say the cutting conditions for rough turning are entered as a speed of 350 SFM, a feed rate of 0.014 IPR, and a depth of cut of 0.125 in.

As you write a program that uses this material, the material is selected during the entry of general information. Later, during your description of machining operations, you specify a rough turning operation. At some point during this operation, the conversational control will display your cutting conditions (as 350 SFM, 0.014 IPR, and 0.125-in depth of cut). At this point, the control is simply reflecting information it found in the material file. If the operator agrees with these cutting conditions, they are simply accepted. However, if the operator does not agree with them, they can be easily changed on this cutting conditions page.

Rough stock size and shape. Most conversational controls want to know as much as possible about the condition of the rough stock being machined. If the rough workpiece is in the form of a standard round or flat bar, simple overall dimensions like bar diameter and length are specified. If the rough workpiece is in the form of an irregular shape, like a casting or forging, many conversational controls allow the operator to specify its size-related qualities.

This information is needed for two reasons. First, given this stock size information, the conversational control can automatically scale the size of the workpiece drawing to fit nicely on the display screen. (Of course the finished workpiece must fit inside the rough stock.) Second, and more importantly, once the finished workpiece shape is described, the control will know how much rough stock is to be removed from the workpiece in each operation.

Location of program zero. Though this specification may be somewhat transparent to the operator on some conversational controls, many require the operator to locate the program zero point. If this is required, it is done early in the conversational programming process, during the general information step. One conversational turning-center control, for example, assumes the program zero point in X (diameter axis) to be the center of the workpiece in X (as it always will). But for the Z axis (left-to-right axis) the operator must tell the control the location of program zero. This is easily done by specifying how much rough stock is to be removed from the workpiece during the machining operations.

Once program zero is specified (either explicitly or implicitly), all dimensions used during the drawing of the workpiece will be taken from this location.

Define finished workpiece geometry. The second step to inputting conversational programs is to define the shape of the workpiece to be machined. Though this step varies dramatically from one conversational control to the next, most conversational controls utilize geometry elements to represent the actual surfaces of the workpiece. For example, lines or arrows are used to represent straight surfaces like chamfers and angles. Circles or arcs are used to represent radii on the workpiece. Other element representations like R for radius, C for chamfer, T for thread, and G for groove may be used for easy input of common shapes needed for workpiece definition.

Generally speaking, the method of geometry input used by most conversational controls allows the operator to work directly from the workpiece drawing. In most cases, the operator will have absolutely no need to make manual arithmetic calculations. As long as the workpiece can be made from the drawing (no errors in the drawing), no math should be required.

Figure 1.11 shows an example of this step for one popular conversational turning-center control. Notice that this drawing would require a manual programmer to make several math calculations, some even involving trigonometry. For a conversational programmer, no math will be required. Notice the series of *element symbols* next to the drawing. Each element symbol tells the conversational control about one surface or characteristic of the finished workpiece.

The list of element symbols in Fig. 1.11 is not complete. Though the skeleton of the workpiece can be pictured by looking at the series of element symbols, the specific data about each surface is not shown. While conversational controls vary dramatically with regard to the exact steps required for workpiece definition, here we give the details of what the operator must input for one popular conversational turn-

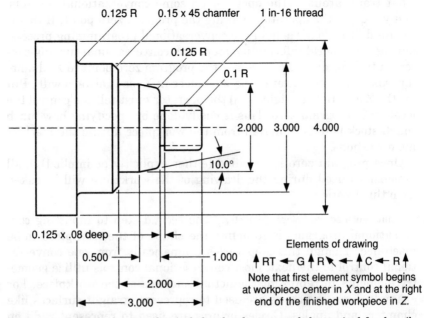

Figure 1.11 Example drawing that includes the element symbols required for describing the workpiece to one popular conversational turning center control.

ing-center control. Note that most conversational controls will prompt the operator for answers at all times. Only what is shown after the colon (:) would be typed by the operator. What is to the left of the colon (:) represents the control prompt.

First arrow up

Start point X: 0

Since it is the first element symbol, the control will first ask the programmer for its starting point. For this example, the first arrow up begins at X0.

Start point Z: 0

End point X: 1.0

First R (radius)

Radius size: .100

T (thread)

Next or last element: Next

For threading, this particular conversational control places a thread on another element. For the next, or last, element question, the control is asking the operator whether the thread is to be placed on the next element (the arrow to the left) or the last element (the chamfer).

Length: .95

Pitch: .0625

Number of starts: 1

Most conversational turning-center controls allow multiple-start threading. The answer to this question tells the control how many starts the thread has.

Depth: .043

Most conversational controls will automatically calculate the depth of the thread based on the pitch.

First arrow to the left

Z end point: 1.0

G (groove)

Next or last element: Last

Like a thread, a groove is placed on another element. This question is asking if the groove is to be placed on the next (line up) or the last (line to the left) element.

Width: .125

Depth: .08

Corner chamfers: No

For grooving, most conversational turning-center controls allow the operator to specify that chamfers or radii are to be included on each corner of the groove.

Second arrow up

End point X: Unknown

For this element, the ending point cannot be known without the operator doing some math. Most conversational controls allow the operator to leave blank those entries that are unknown. As long as the workpiece can be made from the print (no print errors), the control will automatically calculate the unknowns.

Second R (radius)

Radius size: .125

First arrow up and to the left

End point X: 2.0

End point Z: 1.5

Angle: 10.

Second arrow to the left

End point Z: 2.0

Third arrow up

End point X: 3.0

First C (chamfer)

Chamfer size: .150

Third arrow to the left

End point Z: 3.0

Fourth arrow up

End point X: 4.0

Though this example may take some study, notice how easy it is to respond to the control's prompts. The operator can easily do so by using only the print as a guideline. No math was required (even for the tapered surface), since the control allows angular dimensions as well as coordinate values.

Keep in mind that this is only an example of how one popular turning-center control will ask for this data. Our intention is to show you how easy and math-free it can be to program. By no means do we wish to imply that this is the only way to input workpiece geometry. You must be prepared for variations. But if you understand the points made in this section, you should be able to see the power an operator has for defining workpiece geometry when conversational programming is used.

Define machining operations. The third and last step to conversational programming is to tell the control how to machine the workpiece. Before beginning this step, the operator *must* have a clear idea about the machining process to be used. At the very least, the operator should develop a sequence of operations that will be used to machine the workpiece.

For example, in our previous turning-center example, the operator may decide to use this process:

No.	Operation	Tool	Station
1	Rough face	80° diamond	1
2	Rough turn	80° diamond	1
3	Finish face	55° diamond	2
4	Finish turn	55° diamond	2
5	Plunge groove	0.125-in-wide grooving tool	3
6	Thread	60° threading tool	4

With this form completed, the operator will have a clear understanding of how this workpiece is to be machined. While some experienced conversational programmers may be able to get away without this form, the beginner is almost doomed without it for complicated processes.

Admittedly, the process form has nothing at all to do with actual conversational programming techniques. It is simply for organizational purposes. Now let's look at several specific conversational functions that help the operator define each machining operation.

Select the operation type. The first thing the operator will do for each machining operation is to select the machining operation to be performed. Most conversational controls will display a list of possible machining operations and the operator will simply choose one.

For example, a turning-center conversational control will display a list of operations including center drill, drill, rough turn, finish turn, rough bore, finish bore, groove, and thread (among others). A machining-center conversational control will display a list of operations including center drill, drill, tap, ream, bore, counterbore, rough mill, finish mill, and face mill (among others). It will be up to the operator to simply choose the operation currently desired.

Select the cutting tool. As mentioned earlier, most conversational controls incorporate a tool file. The user permanently stores tool descriptions into this tool file for use during programming. During the machining operations step of conversational programming, the operator simply specifies which tool is being used. Some conversational controls will actually display a list for the operator to choose from. Others require that the operator type in an identification number for the tool.

Once the conversational control is told the style of cutting tool to be used, most will display data about the tool so that the operator can confirm that the tool selection was made correctly.

Check the cutting conditions. Based on the material being machined, the machining operation to be performed, and the tool to be used, the control will display a set of cutting conditions to be used for the operation. Most of this cutting condition information comes from the material file. For example, most conversational controls display the cutting speed, feed rate, depth of cut, rapid approach distance, and any other pertinent information relative to the machining operation.

If the operator agrees with the cutting conditions generated by the control, these cutting conditions are accepted. However, the operator will have the ability to change any cutting condition parameter, if desired. Say, for example, a rough boring operation is being performed on a turning center. Say the boring bar is very small and weak. On the cutting conditions page, the operator may not agree with the feed rate and depth of cut for this operation since the material file criteria for rough boring may be based on a strong and rigid boring bar. In this case, the operator could easily modify cutting conditions to ensure a safe machining operation.

Specify the stock to be machined. Most conversational controls allow the operator to be quite specific about designating the rough stock to be removed from the workpiece in each operation. Especially with turning-center conversational controls, the operator will have the ability to specify just how machining is to take place. For example, if rough facing the workpiece, the operator will have the ability to specify that only the material on the face of the workpiece is to be removed, and that no material on the outside diameter or in the bore of the workpiece is to be machined at this time.

Creating a CNC program with a conversational control.   Once the conversational program is finished, most conversational controls actually convert the conversational program into a true CNC program. Once this CNC program is created, it will look identical to any manually generated CNC program. It is from this conversationally generated CNC program that production will be run.

Earlier, we said you can think of a conversational CNC control as having two computers (the conversational computer and the CNC computer) that share the same display screen and keyboard. When the conversational control creates the CNC program, it is also *downloading,* or transferring the program to the CNC-related computer of the control. During this program transfer, most conversational controls also give the operator an important program verification function called *tool-path display.*

Tool-path display.   One of the major advantages of conversational controls is that they are graphic and show the operator many things about the program being prepared. One of the most important graphic features is the tool-path display. This function allows the operator to visually check the series of movements each tool will make during the execution of the program. For most controls, solid plotted lines represent cutting motions and dotted plotted lines represent rapid positioning motions. And since most conversational controls have color display screens, each tool will be plotted by a different color, so it will be easy to distinguish one tool path from another. Figure 1.12 shows the tool-path display screen of a popular conversational control.

During this tool-path display, the operator will even have control of a function that works like single block. If desired, the operator will be able to monitor the program being executed one command at a time. Once the operator monitors the tool path for each tool and is confident that the movements are correct, the operator can rest assured that the CNC program generated will make the machine follow the exact same series of movements.

Graphic image zooming.   Sometimes, for extremely large or long workpieces, it will be difficult to monitor all motions of the tool-path display. As the movements are generated, the various tool lines will run together and make it difficult to truly see what is happening. For this reason, most conversational controls allow the operator to zoom in on an area of the workpiece in question.

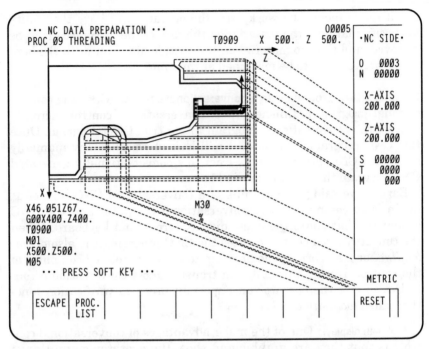

Figure 1.12 Example of tool-path display on a popular conversational control. (*Courtesy GE Fanuc Automation.*)

With the display screen zoomed-in, the tool path can be generated again. This time only the tool path in the zoomed area will be shown, giving the operator a much clearer picture of the area of concern. This zooming procedure can be repeated as many times as necessary, until the operator has no question about what the program is going to do.

Program storage with conversational controls. When conversationally programming, the operator generates two separate programs, the conversational program and the CNC program. Keep in mind that the conversational program generates the CNC program. It does *not* work the other way. That is, a CNC program *cannot* generate a conversational program.

When it comes to CNC program storage, every program-transfer device discussed earlier in this chapter can be used. Indeed, a normal CNC program is involved. However, the conversational program is *not* like a normal CNC program, and in most cases a standard program transfer device *cannot* be used. This is because most conversational control manufacturers use a proprietary (nonstandard) method for conversational program storage. If you wish to save and retrieve conversational programs to an outside device, most conversational control manufacturers require that you purchase the program transfer device from the control manufacturer. As with programs stored within the CNC side memory, a limited number of conversational programs can be stored within the control's memory, on the conversational side.

Computer-aided manufacturing systems

The fourth and last form of program preparation device is the CAM system. To begin our presentation, let's build on what you already know. As presented earlier, a computer-based CNC text editor allows off-line (away from the CNC machine) creation of CNC G code level programs. Once created, these CNC programs can be quickly and easily transferred to the CNC machine. You also know that conversational controls allow a CNC machine operator to prepare a CNC program in a graphic way, utilizing conversational techniques.

You can think of a CAM system as allowing off-line creation of CNC programs by using the conversational style of programming. Many of the graphic-related features of the best conversational CNC controls are incorporated in today's CAM systems. In a way, today's CAM systems are like multipurpose conversational controls, whose purpose it is to prepare CNC programs for *all* of the CNC machines a company owns.

Keep in mind that all forms of computer software application programs vary drastically from one supplier to the next. Word processors, for example, all have the same primary goal: to create documents like letters, invoices, and manuals. However, the actual keystrokes and commands required from the user vary dramatically from one word processor to the next. One may use a mouse while another may not. One may use CONTROL B to specify bolding of text while another may use ALT B instead. And so on.

In the same way, all CAM systems have the same basic goal: to prepare workable CNC programs for CNC equipment. However, they vary dramatically with regard to how they are utilized.

Though the specific software programs for any given computer application may vary, there is sufficient similarity from one to the next to allow a computer user to adapt from one to another with relative ease. CAM systems are no exception to this software phenomenon. Competition among the CAM system suppliers warrants that there will be dramatic differences with regard to how they are used. However, once a programmer masters one CAM system, it will be relatively easy to learn another.

Our intention in this section is *not* to give you every detail of how every CAM system in existence is used. To attempt this could truly fill several volumes. Our goal is simply to relate the basics of how CAM systems are used as program preparation devices. We will show you the best applications for CAM systems as well as some of their most common features. We will also give you a basic understanding of how they are used.

As stated, the goal of all CAM systems is to create a workable CNC program that will machine a workpiece to be run on a CNC machine tool. As with conversational CNC controls, *two* programs will be involved, the CAM system program and the CNC program. Also, as with conversational CNC controls, the CNC program is created by the CAM program. (If you have computer programming experience, you can think of the CAM system program as the source code and the CNC program as the executable file created by the source code.)

Applications for CAM systems. If a company has one CNC machine tool and runs a limited number of simple workpieces, it may be hard-pressed to justify the purchase of *any* CAM system or conversational control. As mentioned much earlier in this chapter, this kind of company will be better off manually programming the small number of different workpieces and entering these programs through the keyboard of the CNC control. Quite possibly, all programs the company runs may even fit in the CNC control's memory; the company may not even need any form of program transfer device (unless it wants to back up the programs stored in the CNC control's memory).

Unfortunately, CAM system salespeople tend to unwittingly mislead CNC users into believing that *every* company that uses CNC needs a CAM system. Indeed, there are even those uninformed people in this industry that believe the *only* way to prepare CNC programs is to use a CAM system.

For the right application, a good CAM system can make life *much* easier for the programmer. But depending on the application, even a very good CAM system could conceivably take longer to program for a given workpiece than using manual (and parametric) programming techniques. In some instances, a good manual programmer can outperform a CAM system programmer by as much as 10 to 1 for simple work.

Keep in mind that no matter how good the application for a CAM system, a basic knowledge of manual CNC programming techniques is *mandatory,* even for a CAM system programmer. We relate this to using an electronic calculator instead of learning how to do arithmetic the old-fashioned (long) way. Truly, the calculator allows its user to make complex calculations *much* faster and easier than doing them longhand. However, most people would agree that it is mandatory that students learn how to do basic arithmetic in a longhand manner *before* an electronic calculator is allowed. Without an understanding of the arithmetic behind each calculation, the electronic calculator's user is working blind and may be willing to believe even the most bizarre calculation results.

In the same way, a CAM system programmer will find it necessary to understand G code level functions on a regular basis. There will be many times during program verification at the CNC machine tool that the programmer will need to make adjustments to the CNC program. At the very least, the programmer will have to make cutting condition adjustments like changing feed rates and spindle speeds. The more a programmer knows about G code level programming, the easier these program modifications will be.

If the CAM system programmer does *not* possess a knowledge of G code level functions and CNC manual programming, *every* modification needed will have to be run through the CAM system. This could take as much as 30 minutes per program modification while the programmer goes back to the CAM system, makes the change, re-creates the CNC program, and transfers it to the CNC machine. Over the life of the CNC machine tool, this could cause countless hours of wasted production time while the machine sits idle, waiting for the CAM system programmer to generate a new CNC program for every little program change.

Truly, the more a CAM system programmer knows about manual programming techniques, the better overall programmer he or she can be. While there are those in the industry who would argue that a CAM system programmer does not need to know manual programming, we strongly disagree with this way of thinking.

As stated, CAM systems tend to be somewhat oversold. In our opinion, there are *only* three good reasons a CAM system should be needed to help you prepare CNC programs. While this may sound like a bold statement, and certainly CAM system salespeople may not agree with the statements made here, you would be surprised at the number of companies that misapply CAM systems. These companies are actually adding to their program preparation time as compared to manual programming methods.

Complex workpieces. As you know, a CNC machine will simply follow a series of point-to-point movement commands. The CNC program contains these point-to-point motions, and the end points for each motion *must* be included within the program. If the program is manually prepared, the programmer must calculate each motion's end point. For simple workpieces that do not include complex geometry, this may be quite easy. But the more complex the geometry of the workpiece, the more difficult it will be for the manual programmer to come up with the needed end points.

Like conversational CNC controls, CAM systems allow the programmer to describe the geometry of a workpiece without the need for longhand math. Generally speaking, the entire shape of even very complex workpieces can be entered into the CAM system without even one calculation being made by the programmer. The more complex the geometry of the workpiece, the more the need for a CAM system.

Multiple CNC machines. One constant thorn in the side of a manual programmer is the amount of command variation from one CNC control to another. For example, one machine may use a G90 to specify the absolute mode of programming. Another may use the G90 to specify a single-pass turning command. Yet another may use G90 for something else altogether.

This is but one of the hundreds of possible variations from one control to another. Given the number of differences from the way one machine is programmed to another, it becomes very difficult for the manual CNC programmer to keep it all straight. CAM systems dramatically reduce the need to program differently from one machine to another. In fact, many CAM systems allow the same (source) program to create a CNC program for any of the CNC machines the company owns. Of course, this assumes the machines in question are all of the same basic style (machining center, turning center, etc.) and all have the capacity to run a given workpiece.

Older CNC (or NC) machines. Some older CNC (or NC) machines are extremely difficult to program manually. They offer limited (if any) canned cycle capabilities to make programming easier and may insist on very cryptic and strict entry formats. Some older NC machines do not even allow the absolute mode of programming, meaning the programmer cannot work from a program zero point. These machines require that incremental programming techniques be used exclusively, and those techniques are very difficult to work with.

For these "dinosaurs" of NC equipment, CAM systems make programming much easier. The CAM system can easily output a CNC program that does not use canned cycles, formats within the strict rules of any CNC or NC control, and outputs an incremental program just as easily as an absolute program.

The two forms of CAM systems. Though some very complex CAM systems require mainframe computers, today's most popular CAM systems run on PC-compatible computers. For this reason, we limit the scope of our presentations to those CAM systems that run on PC-compatible computers. Generally speaking, these CAM systems fall into two distinctly different categories: word address interface systems and graphic interface systems.

Word-address interface CAM systems. Generally speaking, this form of CAM system is older and more difficult to use, though some of today's most powerful CAM systems are still word-address systems. The CNC programming languages APT (automatically programmed tool) and Compact II are examples of word-address CAM systems. With a word-address CAM system, the entire CAM program is written and typed in much the same way a manual CNC program is written and typed. But instead of being typed directly into the CNC control, this source program is typed into the CAM system computer. The computer interprets one command at a time (in much the same way the CNC control interprets one CNC command at a time) and performs the function described by the command. The commands that make up a word-address CAM system program tell the computer how to machine the workpiece. This form of CAM program is composed entirely of text and must be totally completed before the computer can begin to interpret its commands. This means the programmer will have no idea as to whether a mistake has been made until *after* the entire program has been written.

If syntax (format) mistakes are made, the computer will generate an alarm and create a report listing the errors in the program. Most word- address CAM systems will also show a tool path plot (on a separate plotting device) that will show if mistakes have been made within the motions of the program.

Figure 1.13 shows a simple workpiece to be used for an example of how a programmer defines geometry with a word-address system.

Notice that word-address CAM systems require the programmer to label each element of the geometry (with L1, L2, C1, C2, and so on) before the program can be written. Here is a list of commands that define this geometry for one popular word-address CAM system language:

```
L1 = LINE/ 0, 0, 1, 0
L2 = LINE/ 1, 0, 1, 1
L3 = LINE/ 0, 1, 1, 1
L4 = LINE/ 0, 0, 0, 1
C1 = CIRCLE/ XSMALL L2, YLARGE L1, .125
C2 = CIRCLE/ XSMALL L2, YSMALL L3, .125
C3 = CIRCLE/ XLARGE L4, YSMALL L3, .125
C4 = CIRCLE/ XLARGE L4, YLARGE L1, .125
```

As you can see, commands required in a word-address programming language can be cryptic. They require the programmer to know the strict rules of the language. If mistakes are made, either the control will not understand the command and generate an alarm or, worse, the control will misinterpret the command.

What gives these word-address CNC programming languages their

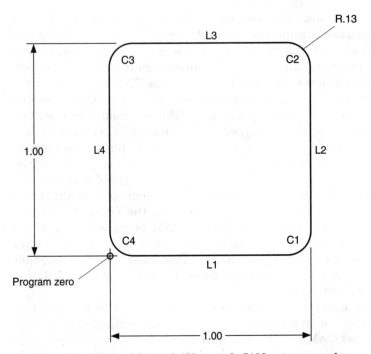

Figure 1.13 Drawing used for word-address style CAM system example.

power is the level of computer programming-language-related features they allow. Most allow the same features as the languages BASIC, FORTRAN, and C. These computer-related features include variable techniques, complex arithmetic, unconditional (GOTO) and conditional (IF) branching, and subroutine (macro) capabilities. These features allow the CNC programmer to define and machine extremely complex shapes such as those required by the aircraft industry for machining airfoils.

Graphic interface CAM systems. Growing in popularity and effectiveness is the graphic interface CAM system. These systems typically run on relatively inexpensive desktop personal computers. Though their cost is relatively low, they boast a great number of powerful features.

These CAM systems are graphically interactive, resembling a conversational CNC control when it comes to being user-friendly. In fact, most highly resemble the best conversational controls. During every step of the programming process, these CAM systems "echo" on the graphic display exactly what the programmer has just done. If a line has just been defined, the CAM system will draw the line on the screen. If a cutting operation has just been described, the CAM system will display exactly what stock has been removed from the workpiece. And so on. Very little is input by the operator that is not immediately shown on the display screen.

This graphic interface makes programming much easier and less error-prone than for word-address CAM systems. Since the programmer will be able to visually check every step of the programming process, he or she will be less likely to miss a mistake.

Since the graphic interface CAM system is more popular than the word-address systems (and growing in popularity every day), the rest of this section will deal with this easier-to-use and more popular CAM system style. Keep in mind that there are *many* different graphic interface CAM systems available from over 100 suppliers in the United States alone. These systems range in price from under $1000 to well over $50,000, and vary dramatically with regard to how programs are created. Our goal will be to acquaint you with the most basic features of graphic interface CAM systems and show you how these features help the programmer create CNC programs.

As with conversational CNC controls, there are three basic steps to preparing any CAM program. And the most basic steps to programming are the same for all graphic interface CAM systems:

1. Supply general information

2. Define the workpiece geometry

3. Describe the series of machining operations

However, there are some points we wish to make about the second step, defining the workpiece geometry, before we describe each step in detail.

CAD versus CAM versus CAD/CAM. The acronym CAD stands for computer-aided design. CAD systems are most often used by design engineering departments to create drawings for the workpieces to be machined by the company. CAD systems allow the drafter to draw the workpiece on the display screen of the computer rather than on a piece of paper. This drawing (in the computer) can be plotted on paper to create the master drawing for the workpiece.

CAD systems allow the drafter to draw on the computer screen almost anything that could be drawn on paper. For example, any line attributes required in an engineering drawing, such as solid lines, hidden lines, centerlines, section lines, and cross hatching, can be easily represented in CAD systems. They also allow a full spectrum of dimensioning and labeling techniques. A good CAD system allows anything the drafter would have done on paper with a pencil to be done within the CAD system.

How CAD helps the manual programmer. If a company owns a

CAD system, it can be used to help with the preparation of a manually written CNC program. As you know, the manual CNC programmer will be very concerned with coming up with end points for each movement the program makes. For complicated contours, this means many arithmetic calculations for the manual programmer.

In this case, the CNC programmer may ask the design engineer to call up the workpiece drawing on the CAD system. From this CAD drawing, the designer or programmer can easily attain the coordinates for any point along the contour. These coordinates can be printed and will be used by the manual programmer during the preparation of the CNC program.

How CAD helps the CAM system programmer.   As stated, the second step to inputting a CAM system program is to define the geometry of the workpiece. Most good graphic-interface CAM systems have elaborate geometry construction definition methods that rival or even improve the definition methods of a good CAD system. This is why most CAM systems are also referred to as CAD/CAM systems. Keep in mind that the CAD portion of most CAM systems is limited to constructing geometry. Very few have the power to perform the line attribute and dimensioning functions required of a true CAD system.

If a company is using a CAD system to prepare its drawings, recreating the drawing in the CAM system is a duplication of effort. For simple drawings, recreating the geometry in the CAM system may be the best method, but for more complicated workpieces, a great deal of time can be saved if the CAD drawing can be used within the CAM system.

For this reason, most graphic interface CAM systems have the ability to read (or import) CAD drawings. In many cases, this can minimize the time it takes to create the CAM system program, since the time it would take the programmer to define the geometry can be eliminated.

However, there are two warnings we would like to give about importing CAD system drawings into the CAM system. The first is related to drawing accuracy.

A design engineer's main concern is creating a drawing within the CAD system that can be easily interpreted by everyone in the company. The drafter will be highly concerned with the appearance and readability of the drawing. There will be times when the design engineer will make an inaccurate drawing just in order to stress an attribute of the workpiece. For example, say the design engineer is drawing a round workpiece that will eventually be machined by a CNC turning center. Say there two diameters that are very close in diameter, one at 2.000-in diameter and another at 2.010-in diameter. In this case, there is only a 0.005-in shoulder from one diameter to the next. If the design engineer creates the drawing in this manner, the 0.005-in face will not even be visible on the master drawing, given the width of the plotting pen.

In this case the design engineer will probably cheat and draw a larger difference in diameters (say 0.050 in instead of 0.005 in) so that the step will be visible on the drawing. The design engineer will correctly dimension the step (as a 0.010-in diameter difference), but the geometry of the drawing in the CAD system will be incorrect.

If this drawing is "imported" to the CAM system, and if the CAM system programmer is not aware of the design engineer's discrepancy in diameters, of course the CNC program created by the CAM system will not be correct.

There are many times a design engineer will cheat in this manner. If the person using the CAD system to draw workpieces is the same person preparing the CNC programs with the CAM system, and remembers the inaccuracy, it can be easily compensated for within the CAM system program. But if more than one person is involved, there must be free communication between the design engineer and the CNC programmer in order to avoid a mistake.

In addition, *much* of the drawing made by the CAD system drawing will be of no value to the CAM system programmer. Again, the design engineer's main goal is to make a good looking, workable drawing. The design engineer will include a great deal of information on the drawing that will be of no value to the CAM system programmer. Much of this nonpertinent information will cause problems for the CNC programmer. For example, the design engineer will include dimensions and notes within the CAD system drawing. These will actually get in the way of the CAM system programmer and will have to be deleted.

Along the same lines, the CAM system programmer will be interested only in the surfaces of the drawing to be machined by the CAM system program. Of course, the CAD drawing will include *all* surfaces of the workpiece. Again, nonpertinent data will have to be deleted.

While most CAD systems allow the user to specify that only a portion of the overall drawing be used by the CAM system, in many cases it will take the CAM system programmer just as long (or longer) to rid the drawing of unwanted geometry as it does to redraw the workpiece geometry from scratch. This is especially true for simple workpieces. While the ability to import CAD drawings into a CAM system is an important one, in our opinion it is somewhat overstressed by CAM system salespeople, given these two major limitations.

How CAM systems are made multipurpose. Earlier we stated that CAM systems are like multipurpose conversational controls. In almost all cases, a conversational control can prepare programs only for the machine tool to which it is attached. Depending on the supplier, CAM systems can create CNC programs for virtually any form of CNC machine tool, including machining centers, turning centers, wire electrical discharge machining (EDM) machines, turret punch presses, laser cutting machines, and any other form of CNC machine tool.

Generally speaking, this is accomplished by making the CAM system modular. By modular, we mean that only the portion of the CAM system needed to create programs for a particular style of CNC machine tool will be loaded into the computer's memory at the time of programming. In fact, most CAM system suppliers will break up the CAM system, selling only those modules a customer has need for. This will reduce the overall price of the CAM system since customers pay only for what they use.

Most CAM systems will use one geometry-creating module for the purpose of defining the workpiece geometry. This geometry module will be used to create the workpiece geometry to be used by every machining module the CAM system allows.

When it comes to machining operations, there is a world of difference between operations that are to be performed on a turning center and those to be done on a CNC machining center. This is true for any two different styles of CNC machines. For this reason, once the geometry is created, the proper machine tool module must be invoked. The CAM system will have a different machine-tool module for each style of CNC machine tool the company owns [machining-center module, turning-center module, wire electric-discharge machining (EDM) module, and so on].

As you can imagine, the CAM system brings a great deal of consistency to the programming department. Since only one style of programming is needed, a programmer can easily prepare programs for a variety of different CNC machines without having to stray from one method of program input. And with a CAM system, it becomes very easy to create one program that is to be used on several similar machines.

How CAM systems handle slight machine differences. Even within one CNC machine group, there will be many differences in programming format from one machine to the next. For example, even among CNC machining centers, there are many formatting differences. One machining center may require an M06 to change tools while another includes the tool change with a simple T word. One may use the letter R to represent the radius in a circular command, while another uses I, J, and K. And these are but two minor examples of the hundreds and hundreds of discrepancies (some much more serious) between CNC machining centers.

Along the same lines, there are many differences of opinion among CNC programmers as to how CNC programs should be formatted. For example, one person may load the first tool at the beginning of a program while another may leave the first tool in the spindle at the end of the program. In order to prepare programs for all varieties of CNC machine tools and to accommodate programmers' individual preferences, CAM systems must be very flexible when it comes to outputting CNC programs.

Most CAM systems incorporate a *postprocessor* that fine-tunes the CNC program to the particular CNC machine tool being programmed. Some older CAM systems require a special postprocessor to be purchased for each CNC machine the company owns. Newer CAM systems let programmers create their own postprocessors from within the CAM system, which minimizes the overall cost of the system.

The postprocessor feature is what allows one CAM system program to be used to create CNC programs for a variety of different CNC machine tools within the same machining module. For example, if a CAM system program exists for one CNC machining center, it can be easily used to create a CNC program for another machining center, assuming that both machines are capable of machining the workpiece.

Steps to CAM system programming. Here we present the most rudimentary steps to CAM system programming. By no means do we wish to imply that this is all there is to it. We will be taking only a cursory look. Truly, the methods of inputting CAM system programs varies dramatically from one system to the next. But, as stated, all require at least these three basic steps:

1. Supply general information
2. Define geometry
3. Describe machining operations

General information. CAM systems require the same kind of general information as conversational CNC controls. Information like program name, workpiece material, the size and shape of the rough stock, the location of program zero, and the machine being programmed are required by most CAM systems during this step.

Most CAM systems will ask the user for this information in menu format. Some will even draw illustrations on the display screen that make it very clear to the programmer as to what is required.

Define geometry. As with a conversational CNC control, the programmer must describe the geometry of the workpiece to be machined. Most CAM systems make this step quite easy, giving the programmer a multitude of geometry definition methods by which to create the geometry.

For simple to moderately complex shapes, three basic geometry elements are involved. They include points, lines, and circles. For each of these elements, there may be as many as 10 to 20 definition methods, depending on the power of the CAM system. Here are a few common definition types:

Points:

X and *Y* coordinate

The intersection of two lines

The intersection or tangency of a line and a circle

The intersection or tangency of two circles

The center of a circle

Lines:

Parallel to an axis

Parallel to another line

Passing through a point at a given angle

Tangent to a circle at a given angle

Circles:

Center point and radius

Tangent to two lines with a given radius

Tangent to two circles with a given radius

Tangent to a line and a circle with a given radius

Passing through a point and tangent to a circle

How these definition methods are displayed by the CAM system and how they are executed by the programmer vary from one CAM system to another. Some make it a little difficult, requiring the programmer to memorize the various definition methods. Others (especially newer CAM systems) make it very easy by displaying a graphic symbol on the screen to represent each definition method. The programmer simply chooses the desired definition method from the choices on the display screen.

Keep in mind that when the programmer describes geometry in the form of lines and circles, most CAM systems will display the entire line or circle. That is, since lines run on infinitely, they will run off the screen in both directions. Circles will be displayed in their entirety.

Older CAM systems required that each geometry element be numbered. For example L1 would represent line 1, C1 would represent circle 1, and so on. These older graphic-interface CAM systems (and word-address systems) required that the previously defined points, lines, and circles be referenced during the machining operation. For example, if a contour was to be milled, every geometry element of the contour had to be referenced during the milling operation in these older systems.

Newer CAM systems will automatically recognize the geometry to be machined; geometry elements no longer have to be referenced during the machining operations description. In fact, there is no longer the need to number each geometry element. To reference an element, the programmer will simply point at the element with the mouse cursor.

However, with newer CAM systems, the programmer must make the geometry on the display screen look precisely like the contour to be machined. The lines that run off the screen in both directions must be trimmed to intersection or tangency points with other lines or circles. In like manner, full circles must be trimmed to intersection or tangency points with other geometry elements.

The methods used by CAM system suppliers to allow trimming of geometry elements vary from one supplier to the next. Most require the programmer to first select the kind of trimming to be done (line trimming, circle trimming, etc.) and then to choose the element to be trimmed on the display screen with the mouse cursor. Finally, the elements representing the tangency or intersection of the line to be trimmed can be chosen with the mouse cursor.

Describe the machining operations.   Once the geometry is described and trimmed, the programmer can enter the third step to CAM programming, to define the machining operations to be performed. Most CAM systems display a list of machining operations pertinent to the machine tool being used to machine the workpiece. If you are programming for a machining center, the operations list will include center drill, drill, tap, ream, bore, contour mill, face mill, and so on. If programming for a turning center, the operations list will include center drill, drill, rough turn, finish turn, rough bore, finish bore, groove, thread, and so on.

The programmer will simply select each operation in sequence of machining. During each operation (as with conversational CNC control programming) the programmer will be prompted to answer questions about the tool being used, cutting conditions, stock to be removed, and any other information pertinent to the machining operation.

We must point out that there are those "multipurpose" CAM systems that are much more helpful with one form of CNC machine than another. The CAM system may have been originally designed around a machining-center application and then "adapted" for use with a turning center (or some other form of CNC equipment). The machining center module of the CAM system may be very powerful, easy to use, and may perform machining operations in the program quite nicely. But when the programmer goes to use another module, the CAM system may not be nearly as helpful. CAM system developers are human. They may have better experience with one form of CNC machine tool than another. When evaluating CAM systems for purchase, *always* ask to see a demonstration for *every* module you will be purchasing. If you don't, you will more than likely see only the most powerful and easiest-to-use module.

Program storage with CAM systems. As with computer-based CNC text editors, CNC programs generated by a CAM system will be saved on the hard drive or floppy diskette. As with conversational CNC controls, two programs will also be of importance: the CNC program which drives the CNC machine and the CAM program which generates the CNC program. The programmer must be concerned with saving the CAM program as well as the CNC program.

When it comes to transferring CNC programs to the CNC machine tool, any of the program transfer devices discussed earlier in this chapter will work.

Keep in mind that most CAM systems come with a CNC text editor which allows CNC programs created by the CAM system to be modified. The text editor can also be used to enter CNC programs from scratch. Of course, every point made earlier about CNC text editors applies to the CNC text editor included with a CAM system.

Conclusion to program preparation devices

We have presented here the basic considerations related to program preparation devices. As stated, each company's particular situation determines which program preparation device is best. We have also given many opinions related to the criteria that determine which device is best. If you are to be involved with the buying decision for one device or another, the basic information in this chapter should help when the time comes to make your decision.

DNC Systems

There is quite a bit of confusion in industry today regarding the term DNC. This confusion is understandable because the acronym DNC is used to stand for two distinctly different CNC functions. One is *distributed* numerical control and the other is *direct* numerical control.

In this section we will clear up the confusion and give applications for both forms of DNC.

Distributed numerical control

As the name implies, distributed numerical control is used to distribute CNC programs to and from CNC machine tools. With this most common form of DNC, programs are transferred to and from the memory of CNC controls in their entirety. When programs are executed at the control, they are run from within the memory of the CNC control. *Any* of the program transfer devices discussed earlier in this chapter could be used as part of a distributed numerical control system. However, every program transfer device discussed to this point has required a manually activated program transmission.

Manually activated distributed numerical control systems. By manually activated, we mean both devices require manual setting before a transmission can be made. With manually activated program transfer devices, the device *receiving* the program must be set first. Then the device sending the program can be activated.

For example, say a company is using a portable floppy-disk drive to transfer programs to and from the CNC control. Say a program is to be transferred *to* the CNC control *from* the portable floppy-disk unit. In this case, the operator *must* get the CNC control ready to receive the program *before* the portable floppy-disk unit can be told to send the program. If you are transferring a program from the CNC control to the floppy-disk unit, the reverse is true.

For any portable program transfer device (notebook computers, floppy-disk units, RAM memory program transfer devices, etc.), this manual activation of each device presents no major time-related problems. Also, since these devices can be used with any number of CNC controls tools (as long as the CNC controls have an RS-232C communications port), they can easily be carried to any CNC machine in the shop to upload and download programs.

However, many companies are using desktop computers as their program transfer device. They may be using the communications portion of a CNC text editor or CAM system to send and receive programs. These devices, of course, are not portable. If program transmissions must be made manually, the distance from the CNC machine to the computer presents its own special problems. Figure 1.14 shows the communications menu of a CNC text editor.

In many companies, the operator simply walks back and forth between the computer and the CNC machine to make a transmission. If transferring a program from the computer to the CNC machine,

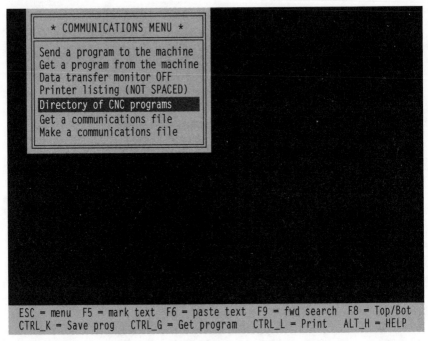

Figure 1.14 Communications menu of a CNC text editor. (*Courtesy Softwerks International.*)

the operator will go to the CNC machine and get it ready to receive the program. Then the operator will walk to the computer and make the command to send the program. If going from the CNC machine to the computer, the operator reverses the procedure.

In some companies, it is a long distance between the CNC machine tool and the desktop computer. Much time can be wasted while walking back and forth between the two devices. Also, some companies do not wish the CNC operator to have access to the computer used for program preparation. These companies may elect to use their internal telephone system to help. When an operator needs a program sent, he or she calls the programmer. While on the phone, the programmer instructs the operator to get the CNC control ready to receive the program. Then the programmer makes the command to send the program from the computer to the machine.

Multiple machines. Another problem that comes up with a manually activated distributed numerical control system has to do with multiple CNC machines. If a central desktop computer is to be used as the program transfer device, and if this computer is to act as a host computer for several CNC machines, the computer must be connected to each CNC machine. This requires, of course, running a separate cable from the computer to each machine.

Most desktop computers have a limitation related to the number of program transfer devices with which they can communicate. As stated, the computer uses its serial communications port through which to transfer programs. Most computers only have one or two serial ports (called COMM1 and COMM2). This means only one or two communications devices can be connected directly to the computer at a time.

When multiple CNC machines are involved, and to save the CNC people from having to constantly unplug cables to connect to different machines, most companies purchase manual switch boxes through which they run the cables. Before a transmission can be made to any one particular machine, this manual switch box *must* be set to the proper position.

For example, say a company has three CNC machine tools: a machining center, a turning center, and a wire EDM machine. One cable has been run from the manual switch box to each machine. On the computer side, the manual switch box is connected to the computer's serial port. In this case, the switch box must have at least three positions. This kind of switch box is called an *A-B-C switch.*

Say the machining center is connected to the A position of the switch box, the turning center to the B position, and the wire EDM machine to the C position. If the machining-center operator needs a program, the switch box must be set to the A position in order for the transmission to be successfully completed.

For a limited number of CNC machine tools, this method of manually transferring programs is not so bad. But as the number of machines grows, this form of distributed numerical control system becomes more and more cumbersome to work with. Also, the amount of program loading time wasted while the operator walks back and forth between the devices increases for each CNC machine added to the system. For these reasons, many companies use a more automatic way of transferring their CNC programs.

Automatically activated distributed numerical control systems.   There are two forms of automatic distributed numerical control systems that allow the CNC machine operator to command program transfers from shop floor near the CNC machine tool. These distributed numerical control systems eliminate the need to manually set each device, thereby avoiding the time-consuming walk between devices. The entire program transfer can be activated from the CNC machine tool. One of the systems actually uses the CNC control panel as a terminal to communicate with the host computer. The other system uses a separate terminal with which to communicate with the serving computer.

Both forms require the same basic CNC machine-tool communication function. Namely, the CNC machine tool must have an RS-232C communications port to allow serial communication.

Using the CNC control panel as the remote terminal. This form of automatic distributed numerical control system is the least expensive. Since this form of automatic distributed numerical control system allows the operator to communicate with the serving computer through the CNC control's keyboard and display screen, no additional hardware is required for each CNC machine tool. This means the cost of the system can be kept quite low. The only limitation of this kind of system is that its only purpose is to transfer CNC programs.

The largest single advantage of any automatic distributed numerical control system is that the operator has the ability to send, retrieve, and sometimes even print the CNC program from the CNC control panel. This, of course, means that there is no longer any need for walking from one device to the other prior to the program's transmission. It also means there is no need for setting the manual switch box if multiple machines are involved. Figure 1.15 shows the display screen of an automatic distributed control system.

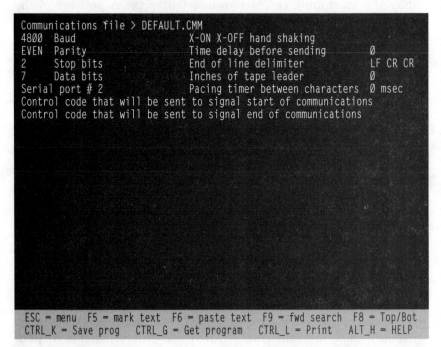

```
Communications file > DEFAULT.CMM
4800  Baud                       X-ON X-OFF hand shaking
EVEN  Parity                     Time delay before sending        0
2     Stop bits                  End of line delimiter            LF CR CR
7     Data bits                  Inches of tape leader            0
Serial port # 2                  Pacing timer between characters  0 msec
Control code that will be sent to signal start of communications
Control code that will be sent to signal end of communications




ESC = menu   F5 = mark text   F6 = paste text   F9 = fwd search   F8 = Top/Bot
CTRL_K = Save prog   CTRL_G = Get program   CTRL_L = Print   ALT_H = HELP
```

Figure 1.15 Display screen of a popular automatic distributed numerical control system. This screen is displayed when the system is waiting for a transmission to be requested. (*Courtesy Softwerks International.*)

Though there are exceptions, this kind of system usually requires that one personal computer be dedicated to the distributed numerical control system. That is, it cannot be used for any other purpose except to support program transfer. This computer will have limited system requirements, since the distributed numerical control system will not tax the computer to any great extent. Just about the only major system requirement is that a hard drive large enough to store all of the CNC programs the company uses must be available. The computer will be left on during all company working hours, running the distributed numerical control software, constantly waiting for a program transfer request.

Instead of a manual switch box, this kind of distributed numerical control system uses an automatic switch box. The automatic switch box is usually supplied by the manufacturer of the distributed numerical control system and can be designed to accommodate as many CNC machines as the company owns.

When the system is up and running, the automatic switch box will be constantly scanning the various machines, looking for a transmission to come through. As soon as an operator at any CNC machine makes a command to send or retrieve a CNC program, the switch box locks in to that particular machine until the entire transmission is completed.

This kind of system usually requires that the CNC operator create a control program that tells the direct numerical control system what to do. This control program tells the system which machine is involved, the kind of action to be taken (send, receive, or print), and the program name to be transferred.

Though these control programs vary from one system to another, here is an example of how one popular automatic distributed numerical control system's control program works. Notice that, since *all* CNC machines can accept N words as sequence numbers, the N word is used as a code word to pass information to the distributed numerical control system.

```
O0001 (Program number for control program)
N0018 (Machine number involved with transmission)
N0001 (Type of action to be taken; 1 = send program)
N0124 (Program number to be sent)
M30 (End of controlling program)
```

With this particular controlling program, the first command tells the system what machine is involved with the command (machine number 18 in this case). The second command tells the system what action to take. (For this particular system, N0001 tells the system to send a program, N0002 tells the system to retrieve a program, and

N0003 tells the system to print a program.) The third command tells the system what program number is involved with the transmission. After this program is sent to the distributed numerical control system, the system will know what to do.

Here is an example. Say the operator wishes to get a program from the host computer. Say the operator is working on machine number 6, and the name of the program the operator wishes to receive is CNC program number 1001. Here is the controlling program the operator would use:

```
O0001 (Program number for control program)
N0006 (Machine number involved with transmission)
N0002 (Type of action to be taken; 2 = retrieve program)
N1001 (Program number to be sent)
M30 (End of controlling program)
```

When ready, the operator connects this machine to the host computer (through the automatic switch box). After requesting a program to be sent to the CNC machine, and once the control program has been sent, the operator will immediately get the control ready to receive a program. As soon as the automatic switch box senses that a transmission is being made, it locks in to this machine for the balance of the action being taken. The host computer receives the program, and the distributed numerical control system now interprets the command program. In this case, the system will know to send program number 1001 to machine number 6. After a short pause (long enough to allow the operator to get the machine ready to receive), the host computer will automatically send program number 1001. For short- to intermediate-length programs, all of this can be accomplished in under 2 minutes! This represents a tremendous savings over any form of manual program transfer device.

To continue with this example, say after the transmission the operator runs the program (number 1001) to machine workpieces. Say that during this program's verification, some modifications are made. The CNC program in the control's memory no longer matches the program in the host computer. After the program is verified, the operator can just as easily send this program back to the host computer for permanent storage. If desired, the program can even be printed so that the operator can have a revised program listing (hard copy). Here is the control program to send the program back to the host computer [note that only one word was changed (the type of action being taken)]:

```
O0001 (Program number for control program)
N0006 (Machine number involved with transmission)
```

N0001 (Type of action to be taken, 1 = send program)
N1001 (Program number to be sent)
M30 (End of controlling program)

After sending this control program to the host computer, the operator would wait a few seconds to allow the host computer to evaluate the control program and get ready to receive program number 1001. Then the operator would send program number 1001 to the host computer.

To print program number 1001, here is the controlling program:

O0001 (Program number for control program)
N0006 (Machine number involved with transmission)
N0003 (Type of action to be taken; 3 = print program)
N1001 (Program number to be sent)
M30 (End of controlling program)

Of course, the operator would have to go to the printer to get the program listing. Some companies even keep a printer centrally located in the shop to keep operators from having to stray too far from their station.

As you can see, the control program can be left in the CNC control's memory and kept available for quick and easy modification. Whenever the operator wishes to perform an action with the distributed numerical control system, this program is easily called up, modified, and sent to the host computer.

Using a separate remote terminal. This form of automatic distributed numerical control system is *much* more expensive, simply because more hardware is involved. With this system, each CNC machine requires its own terminal to communicate with the host computer. These remote terminals range in price from $1000 to well over $2000 each. Keep in mind that when you use this form of automatic distributed numerical control system, *two* actual transmissions must be made—one between the serving computer and the remote terminal and the other between the remote terminal and the CNC machine tool.

There are three possible advantages for this form of automatic distributed numerical control system. First, since a separate terminal is involved, you can modify the CNC program being transferred without having to do it at the CNC control panel. This means the operator can be altering one program while another is running. (Note that some CNC controls allow this as well if the feature *background edit* is available.)

Second, this form of automatic distributed numerical control system usually requires a host computer with more capability than a simple personal computer. Usually a more expensive mainframe computer is required. This more powerful computer will usually have multitasking capabilities; that is, the serving computer will have the ability to serve more than one CNC machine tool at a time. This means the system can double as a *direct* numerical control system. More on direct numerical control a little later.

Third, since the remote terminal can take the form of a personal computer or workstation (with a full-color graphic display screen), many companies incorporate more data transfer into the system than simply communicating CNC programs. They allow the operator to view the workpiece drawing, process list, gauging list, and any other pertinent information related to the job right at the workstation of the distributed numerical control system. This eliminates the need for a great deal of paperwork. However, note that automatic distributed numerical control systems with this capability are very expensive indeed.

Management of the system. Any form of automatic distributed numerical control system must be well-managed. The sheer number of CNC programs a company uses may be enough to make a programmer's head spin. Say, for example, a manufacturing company has 10 CNC machines and each machine runs about 200 repeat jobs. In this case, 2000 existing programs would have to be on the hard drive of the host computer. This is not to mention the additional programs that must be added on a daily basis.

Keeping track of these programs can be difficult and confusing. For this reason, most automatic distributed numerical control systems offer several features to help with management.

First, most allow different subdirectories of the computer's hard drive to be used for the programs involved with each CNC machine tool. In the above example (10 machines with 200 programs each), this would mean only 200 programs in any one subdirectory on the hard drive.

Second, most automatic distributed numerical control systems also provide organized and easy-to-read printed listings of all programs. These listings allow the manager to place a short description alongside each program on the list. This description can include part number, customer name, process sheet number, or any other information pertinent to the program. The list can be given to the operator of each machine to make it easy to tell what programs are in the system as well as what each program does.

Third, most automatic distributed numerical control systems provide an easy way to get programs into and out of the system. Keep in mind that, in most cases, the computer used for the distributed numerical control system will *not* be the computer that is used to prepare the programs. For this reason, CNC programs must be brought into the computer used for distributed numerical control. Usually, floppy diskettes are used for this purpose. The distributed numerical control system will easily read the programs on the floppy drive into the system. After this is done, an updated report of the directory of programs in the system can be generated.

Fourth, one side benefit of automatic distributed numerical control systems is that the system has enough information to help track which machines are calling up programs for production control purposes. With this information, the system manager can easily tell how many different setups have been recently run on each CNC machine tool in the company.

RS-232C protocol. As you have seen so far, almost all forms of distributed numerical control systems, as well as simple program transfer devices, rely on the standards of RS-232C in order to communicate with each other. There are many factors of RS-232C communications that must be properly configured on both communicating devices in order for transmissions to be successfully made. Here we present the factors. These factors, taken together, are often referred to as the *protocol* of RS-232C. Though the information about to be presented may be a little on the technical side, and once program transfers are being successfuly made there is little need for this information, we wish to include this presentation to reinforce your understanding of DNC systems and program transfer devices. Also, if you are going to make your own simple manual distributed numerical control system as discussed earlier, you will need to understand the factors discussed here.

As stated, these factors work together to allow transmissions. If any one factor is not correctly set on one device or the other, the transmission cannot occur. Figure 1.16 shows the RS-232C–related data of one popular distributed numerical control system.

Handshaking. Each program transfer device will have a buffer storage area in which portions of the program are temporarily stored. Once this buffer is full (or close to full), the program transfer device will take the information in the buffer and store it permanently within the device (on the floppy drive of a portable floppy unit, on the hard drive of the computer, in the memory of a CNC control, etc.).

Keep in mind that any program transfer device can only do one thing at a time. As this information is being permanently stored within the device, the transmission *must* halt temporarily. For this reason, when the buffer is close to full, the receiving device will send out a signal to the sending device, telling it to pause temporarily

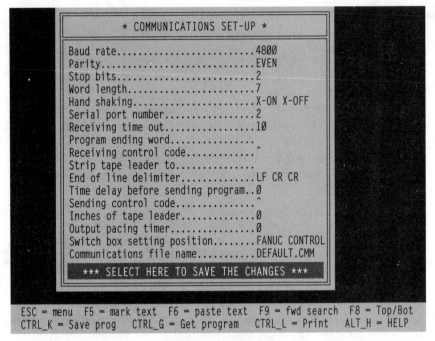

```
          * COMMUNICATIONS SET-UP *

    Baud rate........................4800
    Parity...........................EVEN
    Stop bits........................2
    Word length......................7
    Hand shaking.....................X-ON X-OFF
    Serial port number...............2
    Receiving time out...............10
    Program ending word..............
    Receiving control code...........^
    Strip tape leader to.............
    End of line delimiter............LF CR CR
    Time delay before sending program..0
    Sending control code.............^
    Inches of tape leader............0
    Output pacing timer..............0
    Switch box setting position.......FANUC CONTROL
    Communications file name..........DEFAULT.CMM

    *** SELECT HERE TO SAVE THE CHANGES ***
```

```
ESC = menu   F5 = mark text   F6 = paste text   F9 = fwd search   F8 = Top/Bot
CTRL_K = Save prog   CTRL_G = Get program   CTRL_L = Print   ALT_H = HELP
```

Figure 1.16 Communications menu showing specific parameters of RS-232-C. (*Courtesy Softwerks International.*)

until the information in the buffer is permanently stored. When this is finished, the receiving device sends another signal telling the sending device to continue with the transmission. Depending on the devices involved and the speed of the transmission, this starting and stopping of the transmission may occur many times during the transmission of even a very short CNC program. The process of starting and stopping transmissions for the purpose of emptying the temporary storage buffer is called *handshaking*.

There are two forms of handshaking: hard-wire handshaking and software handshaking. The form of handshaking determines the cable configuration required between the two communicating devices. Generally speaking, the older and less popular form of handshaking is hard-wire handshaking. For this handshaking style, two wires within the cable itself are used to send the stop and start signals.

The newer and more popular form of handshaking is software handshaking. With this form of handshaking, the same two wires used to send and receive the program are also used for handshaking purposes.

Keep in mind that both communicating devices *must* support the same form of handshaking before communication can occur. If one supports only software handshaking while the other supports only hard-wire handshaking, the two devices cannot communicate with each other. Fortunately, newer devices support both handshaking types. They can be used with newer as well as older RS-232C communications devices.

Connectors and cables. Almost all CNC controls manufactured today come with an RS-232C communications port. The RS-232C port on the CNC machine itself is almost always a 25-pin connector (called a *DB25 pin connector*). Most computers made today also utilize the same 25-pin connector for their serial port. However, another connector style that is becoming more and more popular for use with computers (especially laptop and notebook computers) has only nine pins.

In order to correctly make a cable for RS-232C communications, you must know the type of connectors required for the program transfer devices. You must also know the type of handshaking the two devices require. While finding which connector is required is as simple as a visual check, determining which form of handshaking is needed requires a little more digging in the reference manual for the communications devices. Figure 1.17 shows the four possibilities for cable requirements (9- and 25-pin connectors, as well as hard-wire and software handshaking).

Baud rate. This RS-232C parameter controls the rate at which the transmission will be made. Generally speaking, the lower the baud rate, the slower the transmission. The higher the baud rate, the faster the transmission. However, the size of the receiving device's temporary storage buffer also has a great deal to do with how quickly transmissions can be made.

Say, for example, you are transferring programs at a relatively slow baud rate. In this case, the temporary storage buffer may not become full, since the receiving device may have the ability to keep up with the relatively slow transmission, automatically storing information permanently without the need to temporarily halt the transmission with handshaking. However, as the baud rate is increased, the information will flow faster into the receiving device's temporary storage buffer. At some point, the receiving device will need to use handshaking in order to stop the transmission temporarily to empty the temporary storage buffer. The faster the baud rate, the more often handshaking will be required. All receiving devices will eventually reach a saturation point where increasing the baud rate will have little or no influence on the speed of the transmission. The faster the baud rate, the more often the handshaking will stop the transmission.

Baud rate is measured in bits per second (b/s). Standard rates for baud rate include:

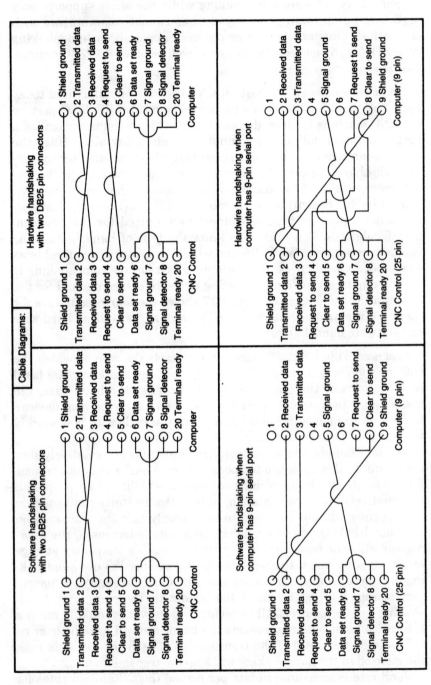

Figure 1.17 CNC–related cables for RS-232C communications.

110 b/s

300 b/s

600 b/s

1200 b/s

2400 b/s

4800 b/s

9600 b/s

19,200 b/s

For almost all CNC program storage devices, 8 bits make up one character of a CNC program. So a baud rate of 2400 b/s would be equal to a transfer rate of about 300 characters of the CNC program per second. At this baud rate, a program that is 5000 characters long would take about 17 seconds to transfer, assuming that the receiving device has an adequate temporary storage buffer and handshaking is not required. Of course, if handshaking is required, this transmission will take longer.

It is imperative that the two communicating devices be set at the same baud rate. If, for example, one device is set to 2400 b/s, so must the other. If the devices are set to different baud rates, the transmission cannot occur. In this case, usually an alarm will sound on one device, the other, or both.

Parity. Parity helps each device check for errors within each character (word) of the transmission. Every word (character) during the transmission *must* be coded in a way that both devices can understand. This coding consists of a series of zeros and ones and is in binary format. If, for example, the word length for each character is 8 bits to the word, both devices will know that *every* word (character) involved with the transmission *must* have 8 bits. If a bit is missing in any word, the receiving device will know that something has been lost during the transmission, and as long as parity is being used, an alarm will be generated.

Most communicating devices give three choices for parity: even, odd, or none. Some serial-type CNC program transfer devices (including some CNC controls) can work *only* with even parity. For this reason, we recommend that you get in the habit of setting all devices to even parity.

Note that if you choose "none" for parity, the device will not be checking for erroneous word makeup during transmissions.

Word length. As mentioned in the discussion of parity, word length is the number of binary digits making up one character. It is the number of zeros and ones included in each word and sets the basic coding structure for the various characters being transferred. Though some devices allow word length to be changed, most CNC program transfer devices require an 8-bit-long word. This assumes the parity bit is included in the designation of the word length. To be more technically correct, let us state it this way: most CNC program transfer devices require a 7-bit word with a parity bit, totaling an 8-bit word.

Stop bits. This RS-232C protocol parameter designates what happens at the end of each word. You can think of it as an end of block (EOB) character for each word during the transmission. As with baud rate and word length, *both* the sending and receiving device must be set with the same number of stop bits. The choices will be 1, 1½, and 2 stop bits. One common number that can be used by many CNC program transfer devices is 2 stop bits.

End-of-line delimiter. As you know, every line of a CNC program must end with an EOB. On the control's display screen, the end of block may be represented as a semicolon (;) or asterisk (*), or with no physical character at all. For program transmissions, these screen representations of the end of block are seldom the actual characters used to end each command.

CNC controls vary dramatically with regard to how each line is ended during a program's transmission. The two most common end-of-line delimiters are the carriage return and the line feed. In some cases, a combination of both is used. When initially setting up communications between two devices, you *must* know what the CNC control will use to end each line. The other device must be tailored accordingly.

Cable length limitations. The cable used for RS-232C (serial) communications is a critical factor in determining the success of the transmission. If you are using a portable device, the cable will be very short (under 10 ft long). In this case, though almost any cable material will work, we recommend that shielded cable be used to avoid interference from electrical devices in the neighboring area.

Note that standard shielded cable can be used to a total length of about 150 ft (though we have heard claims from users who say they have standard shielded cables that are much longer). We recommend, for cables over 150 ft, that (more expensive) low-capacitance shielded cable be used. This kind of cable should suffice for lengths up to at least 250 ft. For cables of extreme length, a line booster must be incorporated in order to enhance the transmission of a weak signal.

For RS-232C communications, 500 ft is about the maximum feasible cable length. If you must run longer cables, you should consider an alternative. A portable program transfer device would be the most inexpensive alternative. However, if you must run extremely long cables, there is another (newer) form of communications called RS-422. The main advantage of this newer communications format is that it allows longer cables to be used.

What does it all mean?   If you currently use program transfer devices, and as long as they are functioning properly, probably you (or someone in your company) have already had to deal with many of the RS-232C protocol issues discussed in this section. However, if you are considering the purchase of your first program transfer device, you are yet to have to deal with RS-232C protocol. If you purchase a device that has been specifically designed for use with CNC, like a tape reader/punch, portable floppy-drive system, a CNC text editor, or a CAM system, the instruction manual should tell you how to handle these parameters of RS-232C protocol. If you have problems, you can contact a person representing the supplier of the device for help.

However, if you attempt to make your own CNC text editor and communications system using a word processor and generic communications software (as discussed earlier), you must be prepared to do a little digging. In the manual for both the CNC control and the communications software, you will find documentation about the RS-232C protocol parameters. If the CNC control manual gives you specific information as to how to set the control, by all means take the advice. If you can find no such information, here is a commonly used way the protocol can be set for both devices:

Baud rate:	2400 b/s
Parity:	even
Word length:	8 (7 data bits and 1 parity bit)
Stop bits:	2
Handshaking:	software
Cable:	As shown in Fig. 1.17
End of line delimiter:	Carriage return

While you may have to modify these settings somewhat, they do make a very good starting point. If you have problems, the alarms you receive from the CNC control or from the other device should point you in the right direction. If you are totally lost, you should be able to contact an applications engineer at your machine-tool builder or control manufacturer to get help.

Direct numerical control systems

The DNC systems discussed thus far have had the sole purpose of transferring programs in their entirety to and from the CNC control. After transferring a program to the CNC control, the program will reside in the memory of the CNC control. It is from within the memory of the control that the program will be executed.

Depending on the application, CNC programs will vary dramatically in length. CNC turning-center programs tend to be quite short, especially if the programmer takes advantage of the special rough turning and boring cycles available from today's CNC control manufacturers. Vertical-machining-center programs tend to be longer, just because of the larger number of tools they can hold within their tool magazines. Horizontal-machining-center programs tend to be longer yet, since they can usually rotate the workpiece held in the setup to machine several surfaces. Since horizontal machining centers usually come with pallet-changing devices, at least two different setups can be run by one program. This means one horizontal-machining-center program could equal many vertical-machining-center programs in length.

The most common forms of machining operations performed on machining centers include drilling, tapping, reaming, boring, face milling, and contour milling. Though a horizontal-machining-center program could become quite lengthy, most programs for horizontal machining centers can easily fit within the control's memory if only basic machining operations are done.

By far, the longest CNC programs are programs for machining centers that generate sculptured surfaces as required in the mold and aircraft industry. These programs, generally prepared by some form of CAM system, generate a three-dimensional (3D) shape by making thousands upon thousands of very tiny motions. Each line of these CNC programs will average approximately 20 characters in length.

For example, say you have a program requiring 5000 commands. Twenty characters times 5000 commands is 100,000 characters or 100K bytes. Since many CNC controls come standard with as little as 32K-byte memory capacity, you can see how a lengthy program can present program storage problems. Remember that *all* CNC controls have a limitation with regard to the maximum size of CNC program that can be stored within the memory of the control. When a program is too long to fit within the control's memory, an alternative to loading the program into the control's memory must be found.

The best alternative is to extend the CNC control's memory capacity. Most CNC control manufacturers can supply extended memory if purchased as an option. If enough extended memory is added to allow the loading of even the company's longest program, there will never be a program loading problem. Keep in mind that no alternative (including direct numerical control) will ever match the convenience and efficiency of being able to run a program in its entirety from within the CNC control. More on why a little later.

Unfortunately, the cost of adding extra memory (depending on the CNC control manufacturer) to a CNC control can be very high. Though prices are falling, and many aftermarket suppliers are now getting into the business of upgrading CNC control memory for a fraction of the cost charged by the control manufacturer (just as is the case with PC-compatible computers), many companies simply cannot afford to purchase extended memory for use within their CNC control. One current CNC control manufacturer, for example, charges over $4500 to add about 32K bytes of internal memory. When you compare this to the PC-compatible computer memory industry, where about $200 will buy 1 megabyte (1M byte) of memory (that's 1000K bytes!), you can see how outrageously expensive it is to add extended memory to a CNC control in today's CNC environment.

A second alternative is to break lengthy programs up into smaller sections. This technique is commonly referred to as *slicing* the CNC program. With this technique, the operator will load the first slice of the program, run it, and then delete it from the control's memory. The second slice will then be loaded, run, and deleted. This will be repeated until the entire program is run.

When lengthy programs are run on a limited basis, or for only one workpiece, this may not be a bad alternative to running the entire program from within the control's memory. Since many CAM systems oriented to the mold industry allow the programmer to easily specify the length of each program slice, creating the numerous programs is not at all difficult. However, as the number of program slices related to the job grows, and as the number of workpieces to be machined in each setup increases, this method becomes more and more cumbersome and error-prone for the CNC machine operator.

The third alternative is direct numerical control. With this form of DNC, the program is *not* run from within the memory of the CNC control. It is run from an outside device, such as a computer or portable floppy-disk drive. In essence, the CNC control will execute the program as if it were being run from the tape reader of the CNC machine. In fact, one popular form of direct numerical control system is attached directly to the tape reader port. But instead of the tape reader feeding the program to the CNC control from tape, the direct numerical control system feeds the program to the CNC control from an outside device. The outside device may be a portable floppy-disk unit, a personal computer, or any other form of program transfer device.

There are some limitations of direct numerical control techniques you must be aware of, and we will present them now. Even with these limitations, however, most CNC people still feel that direct numerical control makes a cost-effective alternative to adding program storage memory to the CNC control in today's CNC environment. This feeling may change as the cost of adding internal memory to the CNC control drops.

Keep in mind that when direct numerical control techniques are used, the user is giving up one of the primary benefits of CNC (over NC) in the first place: the ability to modify programs from the keyboard and display screen of the CNC control. If the program must be modified when direct numerical control techniques are used, the modification must be done off line, away from the CNC control.

For this reason, a computer makes a very good choice for the direct numerical control device. Since the user can use a text editor within the computer, the program can be easily modified. In fact, most computer-based direct numerical control systems come with a CNC text editor. Compare this to a portable floppy-drive system or any other device that does not allow editing. If the program being run must be modified, it must be transferred to a device that does allow programs to be modified. This will waste precious production time.

Another limitation of direct numerical control, as compared to running the program from within the CNC control's memory, has to do with running from the middle of the CNC program. If a tool breaks or dulls in the middle of a very lengthy series of motions, say 20 minutes into the program, the operator will want to restart the new tool from the point where the dull or broken tool left off. If you are running a program from within the CNC control's memory, picking up in the middle of the program for this purpose is relatively easy. But picking up in the middle of a program coming from a direct numerical control device can be much more difficult (if not impossible).

One last limitation of direct numerical control techniques has to do with the transmission of the program itself. When it comes to executing CNC programs, nothing will beat the efficiency and integrity of executing programs from within the CNC control. When executing CNC programs from an outside direct numerical control device, a portion of the program must *first* be transferred to the CNC control before the control can execute it. For programs running at relatively slow feed rates with relatively large axis departures, most CNC controls can easily keep up with the transmission rate. However, as the cutting feed rate is increased, and/or as the departure distance per CNC command becomes smaller, it will become more difficult for the CNC control to keep up with the transmission. In extreme cases, the CNC control will bog down and not be able to run at the desired feed rate.

Just how fast a CNC machine can feed during a direct numerical control transmission can be difficult to determine. Two factors contribute to the maximum feed rate possible. The first factor is the size of the motion within each command. The smaller the average motion per command, the more information the CNC control will have to process, and the slower the maximum feed rate will be.

The second factor is the *clock speed,* or internal processing speed, of the CNC control. Just as in the personal computer industry, the microprocessors equipped within CNC controls vary with regard to how fast they can process information. The faster the processing speed of the control, the faster it can execute CNC commands. If you know the difference between a 386- and 486-class personal computer, you know the kind of performance difference we are talking about.

For speed improvement, just as a personal computer can be equipped with a math coprocessor to improve its performance, so can a CNC control be equipped with a *high-speed buffer* to speed its ability to process CNC programs. Today's highest class of CNC-based microprocessors are in 32-bit controls, and most CNC control manufacturers are trending toward this style of microprocessor. However, some older 16-bit controls still exist. For direct numerical control purposes, when fast feed rates are required in conjunction with small axis departures, the 32-bit control should be your control of preference.

If the CNC machine will be machining a hard material like tool steel at relatively low feed rates [under about 15 inches per minute (IPM) or so], most CNC controls will be able to keep up regardless of how small the motion commands are and no matter what class microprocessor is used within the control. On the other hand, if the CNC machine is going to be machining a softer material, like aluminum, feed rates will be much faster. In this case, the control's processing time will be much more critical. To determine just how fast the machine will be able to move during direct numerical control, the control manufacturer must be contacted. If purchasing a new CNC machine, we also recommend that you talk to someone who currently uses direct numerical control with the control you intend to purchase and can give you the real-world answers to just how fast a machine can go under the influence of direct numerical control.

Keep in mind that this same problem exists even if a program is run from within the CNC control's memory. Though the processing time required to execute each command will be shorter (since no transmission is involved), there will still be a limitation to the maximum feed rate, based on the motion distance of each command and the control's clock speed.

Another transmission-related limitation of direct numerical control is related to the CNC integrity of the program being run. When running a program from within the CNC control's memory, the operator can rest assured that the program will run exactly as entered. However, when running from an outside device, there is always the possibility that the program may be corrupted during the transmission. The most common cause of transmission corruption is an electrical surge in the area of the cable. Welders and EDM equipment are notorious for generating this kind of surge, and your RS-232C cable should be kept a good distance away from them. Though the basic features of RS-232C are designed to detect any corruption of data (with parity checking) and report alarms if they occur, there is still this possibility for corrupted data. Even though corrupted data will be detected and may cause no improper machining motions, it will cause a stoppage in the machining cycle.

The memory capacity advantages of direct numerical control. As mentioned earlier, a CNC user pays a premium for internal CNC control memory. By comparison, if an outside device is used from which to run programs, memory capacity is close to unlimited and the price is very small. For example, say a portable floppy-drive system is used as the direct numerical control system. The floppy-drive system itself will cost about $3000. Say the floppy-drive system uses $3\frac{1}{2}$, 1.44M byte high-density floppy diskettes (about $2 each). This system would allow a program of up to 1.44M bytes, or 1,440,000 characters in length. That's about 12,000 ft of tape! To purchase this amount of memory from many CNC control manufacturers may cost as much as $15,000!

One advantage of PC-compatible computers as direct numerical control systems (over portable floppy-drive systems) is that they can be purchased with hard drives. By today's standards, a "small" hard drive is about 30M bytes. And an entire computer system containing a hard drive with over 200M bytes can be purchased for under $3000. This gives the user 200 million characters of program storage, or 1,666,666 feet of tape!

For extremely long programs (over about 2M bytes), direct numerical control may be the *only* way to run the program in its entirety, since there is a limitation to the maximum amount of storage capacity that can be added to the CNC control itself. Though this varies from one CNC control manufacturer to the next, many cannot supply the control with over about 2M bytes of internal memory.

Though most experienced CNC people would agree that nothing beats the convenience and efficiency of running CNC programs from within the CNC control, it is *much* easier to justify the purchase of a direct numerical control system because it may cost tens of thousands of dollars less than adding an equivalent amount of memory within the CNC control.

How direct numerical control systems work. As stated, when you are using direct numerical control techniques, the CNC program is being run from some kind of outside device, instead of from within the control's memory. For any direct numerical control to work, this outside device must be successfully interfaced with the CNC control. By *interfaced*, we mean the CNC control must be able to function properly with the direct numerical control device. This involves not only being able to receive the program (as would be the case with distributed numerical control); the CNC control also must be able to execute the CNC program *as* it is being received.

There are two common ways to accomplish this interface, and the method chosen depends on the CNC control manufacturer and the age of the CNC control. Some CNC control manufacturers make it easier than others to equip the machine with a direct numerical control system. Generally speaking, newer CNC controls are easier to interface with a direct numerical control system than older ones. Unfortunately, there are those CNC control manufacturers (even today) that make it next to impossible to utilize a direct numerical control system. More on why a little later.

Here are the two ways to interface direct numerical control systems to a CNC machine.

Behind-the-tape-reader systems. The older and more cumbersome method of interfacing a direct numerical control system to the CNC control is to utilize the tape reader port. This form of direct numerical control hookup is called a *behind-the-tape-reader* (BTR) connection, and should be used only as a last resort. Once this kind of direct numerical control system is interfaced, the operator will place the control's mode switch to the *tape mode* position in order to run the program. When the cycle is activated, the control will think the program is coming from the tape reader, when in reality, it is coming from the direct numerical control device.

The connections required for a direct numerical control system interfaced through the tape reader are somewhat difficult to make, and usually require the help of the company supplying the direct numerical control system. Keep in mind that once this form of direct numerical device is installed, it is likely that the machine's tape reader will be disabled, meaning tapes can no longer be used to transfer programs. Though there are exceptions to this statement, you will want to confirm whether you will be able to use the tape reader again *before* you have the direct numerical control system installed. This is why we say the BTR style of direct numerical control system should only be used as a last resort.

Connecting through the communications (RS-232C) port. Newer CNC controls allow the direct numerical control device to be connected to the RS-232C port, meaning they allow the communications port to be used for two purposes. They allow the communications port to be used to simply transfer CNC programs into the control's memory (as is the case with distributed numerical control) as well as to be used to execute in the run mode (as is the case with direct numerical control).

Though controls vary with regard to how this is done, most allow the operator to select the communications port usage by some simple display screen choice. Once the selection is made to use the RS-232C port as a program running port, the control will know to execute the program from the RS-232C port. The actual mode switch selection is usually the *tape* mode. But instead of executing the program from the tape reader, the control will actually execute the program from the RS-232C port. Once the cycle is activated, the control will wait for a program to come through the RS-232C communications port. The operator will then command the direct numerical control device to send the program to the CNC control. As the program comes into the CNC control, it will be executed.

Note that handshaking becomes *much* more important when running in this manner. It is likely that the CNC control will be constantly telling the direct numerical control device to temporarily stop sending data until some lengthy machine motion or miscellaneous function is completed.

This method of connecting the direct numerical control system to the CNC control is preferable to a BTR system for several reasons. First, no machine modifications are required for installation. The direct numerical control system is simply attached to the existing RS-232C port. Second, the installation is much easier, meaning it may not be necessary to have the supplier of the direct numerical control system make the installation. Third, *any* form of program transfer device having RS-232C capabilities can be used as a direct numerical control device. Forth, if the machine has a tape reader, it will not be disabled after the direct numerical control connection.

How to determine if a particular CNC control can accept direct numerical control. As stated, not all CNC controls can handle direct numerical control techniques. Even some brand new, current-model CNC controls being manufactured today do not have this ability. When evaluating a CNC machine and control for purchase, it can be very difficult indeed to determine the control's capability in this regard.

Keep in mind that there is a great deal of confusion regarding the acronym DNC. It can stand for *distributed numerical control* as well as *direct numerical control*. And, unfortunately, most CNC machine-tool salespeople *do not* understand the basic points related to distributed versus direct numerical control we have made in this section. If you simply ask the question, "Does this control have DNC capability?," the machine-tool seller will simply (and probably correctly) answer, "Yes." The control in question may have the ability to send and receive programs through the RS-232C port (as with distributed numerical control), but possibly *not* the ability to actually run a program from an outside device.

If you revise your question to "Does this machine have direct numerical control capabilities?," you're doing better, but you still may not get a correct answer. A machine-tool seller who does not understand the difference between direct and distributed numerical control may still answer "Yes," thinking you mean the ability to simply send and receive CNC programs.

As a machine-tool buyer, how do you confirm whether a particular control has the capabilities you need? The best way is to bypass the machine-tool seller and talk directly to a representative of the *control's* manufacturer.

Even then, be cautious with your questions. Your best question is "Does this CNC control have the ability to actually execute a program and machine a workpiece from an outside device, such as a personal computer?" If the answer is yes, press further. Ask if this is done through the RS-232C communications port or if it requires a BTR connection. Ask if any machine modifications or optional purchases are necessary. If you do not feel that you are getting correct answers, ask for the phone numbers of other companies using the control for direct numerical control purposes. They should be able to tell you what possible surprises you are in for.

The actual criteria that determine whether a CNC control has direct numerical control capability are as follows.

For BTR direct numerical control systems. If the CNC control has a tape reader *and* if programs can be executed from the tape reader mode, it is probable that a behind-the-reader system can be successfully installed. But let us warn you here. Some CNC machines come with tape readers but *cannot* be run from the tape reader mode. The tape reader's sole function is to load a program in its entirety into the memory of the control. Once loaded, the program can be executed *only* from within the control's memory. These CNC controls make it virtually *impossible* to utilize direct numerical control techniques.

For RS-232C communications port connection systems. It can be somewhat more difficult to tell if a control has the ability to handle the direct numerical control connection through the RS-232C communications port. It usually takes talking to a representative from the control manufacturer to find out. As mentioned earlier, it must be made very clear to the control manufacturer representative exactly what is expected, because of the confusion and misconceptions that exist about DNC.

Another way to confirm whether a particular control has direct numerical control capabilities through the communications port is to talk to a supplier of direct numerical control systems. Most aftermarket suppliers of direct numerical control systems have done their homework, and can easily tell you whether the connection is possible. These companies are installing direct numerical control systems on various CNC controls every day.

Unfortunately, many companies buying CNC machines have little or no concern for direct numerical control until after the machine is purchased. Either they do not recognize the potential need for direct numerical control, or they make the assumption that the control has the capability. If your company anticipates the need (or even the potential need) to run extremely long programs, it is mandatory that you check into the control's direct numerical control capabilities *before* the machine is purchased.

Program Verification Systems

CNC machine time is very expensive. In many companies, machine time is actually quoted in dollars per hour. Depending on the type of machine tool involved and the competitiveness of the company, this cost can range from $20 to $200 per hour for a given CNC machine tool.

In all companies, a certain percentage of CNC utilization time is not very productive. A CNC machine tool is not capable of producing workpieces for every minute of every day. Though the importance of this downtime varies from company to company, most consider any time the CNC machine is not actually producing workpieces to be time wasted. One of the most time-consuming nonproductive periods is setup. The longer it takes to make a setup, the more production time is wasted.

When you are running programs that have been previously verified (repeat jobs), setup time is relatively consistent and easy to predict. Setups for repeat jobs require only the time it takes the operator to assemble tools, make the work-holding setup, and load the program. In this case, only a minimum of time (if any) will be wasted while trial-running the program. Though the operator must still exercise a certain degree of caution, since the program has run successfully before, the operator can rest assured that it will run properly again.

On the other hand, when running new programs, several time-consuming verification procedures must be followed in order to ensure the safety of the operator. If verification is done entirely during the CNC machine-tool setup, a great deal of production time can be wasted. In extreme cases and especially for low production quantities, it can actually take longer to verify a program than it takes to run the workpieces!

Companies vary with regard to how far they will go to ensure the correctness of a new CNC program. Some will not do very much at all in this regard, and use the CNC machine as the program verification device. They will perform a machine-lock dry run and several free-flowing dry runs to assure that the motions of the program are correct.

Other companies will do almost *anything* to ensure the correctness of a CNC program *before* it is loaded into the CNC machine. They feel the less time it takes to verify the CNC program at the machine tool itself, the longer the CNC machine will be producing workpieces. Though there are still things that could be wrong with the program even when verification devices are used (and we will discuss them later), a program that has been prechecked on a program verification device will be *much* safer and easier to work with. If an outside verification device is used, less time will be wasted at the CNC machine. In this section, we will discuss several devices used for the purpose of verifying CNC programs.

CNC tool-path plotting devices

There are three types of program verification devices that give the programmer a view of the tool path a program will generate. By seeing the physical series of motions each tool in the CNC program will make, the programmer can easily find the motion errors (if they exist) in a program.

Generally speaking, tool-path plotters will only show the tool motions. Most will not actually draw the workpiece or cutting tool. For this reason, the programmer must be able to visualize the style of tool being used as well as the surface being machined during the plotted motions.

With almost all tool-path plotters, cutting motions (like G01, G02, and G03) will be shown as solid lines, and rapid motions (G00) will be shown as dotted lines. If the device allows colors to be used, usually each tool's path will be represented by a different color.

Plotters. This common form of program verification device plots the tool path by utilizing a series of colored pens and paper. As with most computer hardware, a plotter by itself has no ability to do anything. The plotter must be used in conjunction with some kind of computer software. In most cases, a CAM system of some kind is used to drive the plotter.

A plotter will physically draw the series of motions that a CNC program will generate, utilizing a different colored pen to represent each tool in the program. The programmer can watch the plotting being done to determine the exact order of machining. As with all tool-path

Figure 1.18 Plotter used for generating tool-path plot of CNC programs. (*Courtesy GE Fanuc Automation.*)

plotters, solid lines will be used to represent cutting motions and dotted lines will be used to represent rapid positioning motions. Figure 1.18 shows a common plotter used to plot CNC programs.

When the plotting is finished, the programmer will have a physical drawing on paper of what the program will do. When the programmer is satisfied that the program is correct, this hard copy of the plotted drawing can be included with the other documentation related to the program and will make it very easy for anyone working with the program in the future to easily visualize what the program will do.

As with all tool-path simulators, most plotters can only show the actual tool path that will be generated by the program. That is, nothing about the configuration of the tooling or physical workpiece will be shown during the plot.

Keep in mind that most plotters, if they work in conjunction with a CAM system, will show only a representation of what the CAM system thinks the CNC program will generate. That is, most plotters do not work from the CNC G code level CNC program. If changes are made to the CNC program (without using the CAM system), most plotters cannot show the altered tool path affected by these changes.

With the advent of high-resolution graphics on relatively inexpensive personal computers, and since the computer screen graphics from these computers can be printed to high-resolution printers (like laser printers), plotters are becoming less commonplace (though they are still used extensively by some companies).

The display screen of a CNC control.   Many CNC controls are now coming with tool-path plotting capabilities. This feature lets the programmer or operator view the tool path that a CNC program will generate with the display screen of the CNC control. This allows the operator to visually check the tool motions in much the same way a plotter will. As with all color tool-path plotting devices, most color graphic CNC display screens will represent each tool in a different color. Figure 1.19 shows the display screen of a CNC control with tool-path display.

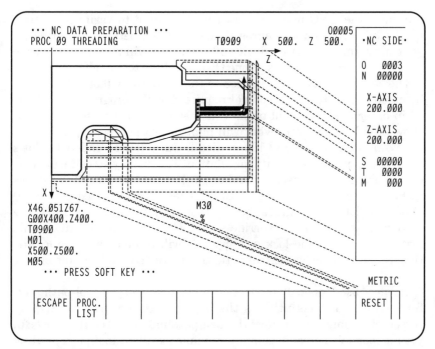

Figure 1.19 CNC control display screen showing tool-path display. (*Courtesy GE Fanuc Automation.*)

One major advantage a CNC-control-based tool-path plotter has over many plotting devices that work in conjunction with a CAM system is that it can plot a CNC program at G code level. The tool-path plotter within a CNC control will show exactly what will happen when the program in the CNC control's memory is executed. If a CNC program is edited, the tool-path plotter within the control will show a tool path including the corrections.

Another advantage of a CNC-control-based tool-path plotter is that it will support all functions of the control. Since the tool-path plotter is developed by the CNC control manufacturer, all functions of the control can be plotted. For example, most CNC controls have a series of canned cycles and special programming features designed to make programming simpler. Some even have very elaborate parametric programming features. The tool-path plotter internal to the CNC control will be able to plot motions generated by these higher-level CNC commands.

While tool-path plotting is an extremely good feature to have within a CNC control, there are two possible limitations. Though this form of tool-path plotting system is much better than having no form of program verification device at all, it defeats one of the primary purposes of having a program verification device in the first place: to minimize the CNC machine downtime required to verify programs. Most CNC people would agree that it is much better to incorporate your program verification device off line. This will minimize the time it takes to verify programs at the CNC machine tool itself.

The second limitation is related to the first. Many tool-path potters included within the CNC control require that the program be activated in order to show a tool path. This means the tool-path plotter will show a real-time view of what will happen as the program is activated. While this may sound like a nice feature, it tremendously slows down the process of tool-path simulation. The display screen will show each movement the machine makes as the machine actually executes it. For example, if a motion of 10 in is required at a feed rate of 10 IPM it will take 1 minute for the tool-path simulator to show the motion (if the feed-rate override switch is set to 100 percent). While features like machine lock and dry run will dramatically reduce the time required to show the tool path, this is wasted production time.

Computer-based tool-path plotters.   The last category of tool-path plotters we will discuss utilizes the high-resolution graphic display screen of a computer to show the tool-path motions. Today's powerful yet inexpensive personal computers make excellent tool-path plotting devices. Tool-path plotters based on computers fall into two distinctly different categories.

CAM-system-based tool-path plotters.   By far, this is the most common form of tool-path plotter. In conjunction with writing almost any CAM program, the programmer will be able to see a tool path displayed on the display screen of the computer.

Most CAM-system-based tool-path plotting devices have one major limitation. Most can only plot a representation of what the CAM system thinks the CNC program will do. Most can only show this tool path *as* the CNC program is being generated. The CNC program is *not* being used to actually generate the tool path.

Unfortunately, if changes must be made to the CNC program at G code level, they cannot be verified by a CAM-system-based tool-path plotter that works in this manner. While some CAM systems do include a tool-path plotter that plots a CNC program at G code level, the overwhelming majority do not. This obviously means that only CAM-system-generated programs can be plotted. This type of tool-path plotter cannot show a tool-path display for a CNC program that has been manually prepared.

G code level tool-path plotters.   Since this form of tool-path plotter works with programs at G code level, *any* CNC program can be plotted. This includes CNC programs that have been prepared manually

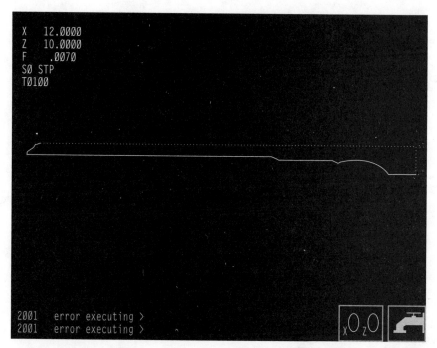

Figure 1.20 Display screen of a popular computer-based tool-path plotter. (*Courtesy Softwerks International.*)

as well as CAM-system-generated CNC programs. Most G code level tool-path plotters even include basic CNC text editors to allow quick and easy corrections to be made to the CNC program if errors are found during the tool-path plotting. Figure 1.20 shows an example of a computer-based tool-path plotter that works at G code level.

This form of tool-path plotter is very similar to the kind based within the CNC control in that it must be able to support the various high-level programming functions of the control in order to show the tool path correctly. G code level tool-path plotters vary with regard to this capability. Some support only the most basic motion codes like G00, G01, G02, G03, and so on. With them, there may be certain functions of the CNC program (like canned cycles) that cannot be shown. Others have elaborate interpreters that are specifically designed for a particular CNC control. These will have the same plotting capabilities as CNC-control-based tool-path plotters.

Simulators

This form of program verification device attempts to actually replicate what will happen as a workpiece is being machined. As with tool-path plotters, simulators can be CNC-control-based as well as computer-based. The same production-time-related points made earlier apply to simulators. Also, as with tool-path plotters, simulators vary with regard to whether they work at G code level or at CAM system level. Those that work at CAM system level can display only CAM-system-generated programs. Those that work with G code level programs can also display CNC programs manually.

Along with the tool-path display, simulators actually show the workpiece being machined as well as the cutting tools doing the machining. Some even show the work-holding devices such as the chuck of a turning center or the vise of a machining center. For machining-center applications, some also allow obstructions like clamps to be shown.

When a simulation is performed, the programmer will be able to see precisely how each tool will machine the workpiece. As each tool machines, most simulators will make the material being machined actually disappear from the display screen. It will be almost as if the programmer is watching a videotape of the machining cycle. Figure 1.21 shows the display screen of a computer-based turning-center simulator.

Unfortunately, there is a price to be paid in time and effort in order

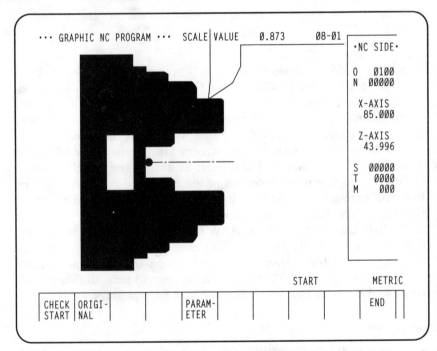

Figure 1.21 Display screen of CNC control-based turning-center simulator. (*Courtesy GE Fanuc Automation.*)

to perform a simulation. Most simulation devices require *much* more work on the programmer's part than simple tool-path plotters. For example, each cutting tool must be described in order for the simulator to correctly display the tool. The stock size and configuration must be described in order for the simulator to display the workpiece in its rough state. If the work-holding device is to be shown, it must also be described. So must any obstructions like clamps and bump stops.

Depending on how difficult it is for the programmer to set up the simulation, there are those in the industry who feel that the simulation of a CNC program may not be worth the effort. In most cases, a good CNC programmer will be able to see all that is needed during a simple tool-path plot. The extra visual checking offered by the simulator may not be sufficient to warrant the extra effort required to set up the simulation for relatively simple applications.

One time when this kind of simulation is *extremely* helpful is for four-axis turning-center applications. If the programmer is working with a turning center that has two turrets, and if the machine has the capability of machining with two tools on the workpiece at the same time, it is mandatory that the two turrets be properly timed. If they are not, the results could be disastrous. For this reason, most CNC people agree that the four-axis turning center is among the most difficult CNC machines for which to verify programs. For this kind of CNC machine, the time required to set up the simulation is always time well-spent, since the simulation will show any timing problems.

What program verification devices cannot show

Though program verification devices will help immensely and should be used if available, every program verification device will have limitations with regard to how perfectly it allows the program to be verified. No matter how sure the programmer is that a given program is correct, there are certain mistakes the CNC machine-tool operator can make during setup that will cause even a perfectly prepared CNC program to fail. This means a certain amount of caution *must* still be exercised by the operator when the program is run for the first time, even when program verification devices are used. There are four possible mistakes that can be made which no program verification device can adequately show.

The designation of program zero. By one means or another, the CNC operator *must* tell the CNC control the location of the program zero point. While some CNC controls make this task easier than others, it is something *all* CNC operators must deal with for every setup they make. If the specification of program zero is incorrectly made, the

CNC control will not truly know where program zero is located, and the motions the machine will follow will not be correct.

This kind of mistake is impossible for any program verification device to adequately display. While some do allow the program zero point to be specified within the program verification device, there is nothing to guarantee that the CNC machine operator will correctly repeat this entry.

The specification of tool length and dimensional tool offsets. In the same way, no CNC program verification device can adequately display the tool length compensation function of a CNC machining center or the influence of tool offsets on a turning center. Again, while some do allow the user to enter these values before the display of the program verification, nothing will assure that the operator will correctly enter these values at the CNC machine tool when it comes time to run the program.

The specification of tool radius offsets. In like manner, no CNC program verification device can adequately display the effects of tool radius compensation in a way that truly guarantees that the program will run flawlessly. If the CNC machine operator forgets to enter the offset values, or enters them incorrectly, the program will still fail to machine the workpiece correctly.

The cutting conditions (speeds and feed rates) related to the job. No CNC program verification device will confirm that the cutting conditions related to a CNC program are correct. While they may allow you to see the depth of cut for each machining operation, they cannot truly tell you whether the machining operation will be correctly performed without some kind of machining practice problem.

Likewise, neither will the CNC program preparation device show whether the program's speeds and feed rates are correct. During the running of the first workpiece, the operator must be very cautious in this regard.

Final note on the limitations of program verification devices. For these reasons, the CNC machine-tool operator must still be relatively cautious whenever verifying new programs. For the most part, if a program verification device is incorporated, the program will be *much* easier to verify. Single block can be used during each tool's approach to the workpiece. If a tool correctly approaches the workpiece and stops the proper clearance distance away from the surface to machine, the operator can rest assured that the tool will move through its motions correctly. At this point, just about the only possible problem with the tool will be with the cutting conditions used for machining.

Other Computer Software Related to CNC

There are two other computer-based software categories we wish to mention. While these software applications have little to do with the actual preparation or verification of a CNC program, they are things a CNC person should be aware of.

Cost-estimating software

Companies that do work for other companies (like job shops) must have a way of properly quoting their jobs. Most quote their jobs on each machine's "shop rate." The shop rate is the amount in dollars per hour a company will charge for the use of any one machine.

In order for this kind of company to accurately quote a job, it must know how long it will take to machine each workpiece. Once they know this cycle time, they can easily determine how long the machine will be in production to run the workpieces. They must also consider the programming time, setup time, and tooling costs. To do all of this manually can be tedious, time-consuming, and error-prone.

Cost-estimating software streamlines the process of quoting jobs. It gives the person doing the quoting a way of quickly and easily estimating the costs related to a job. Most incorporate elaborate material databases which automatically calculate cutting conditions like speeds and feeds. Most allow machine-related information like maximum RPM and horsepower to be easily input. Most allow machining operations (like drilling, tapping, milling, and turning) to be easily input.

Once the estimate is completed, a printout of the estimate can be generated. This printout will include all operations related to the job as well as the cutting conditions related to each operation. If the company gets the job, this information can be used by the programmer in order to assure that the program is prepared in the same way as the estimate.

Cutting conditions software

All CNC users are highly concerned with applying efficient cutting conditions to the operations they perform. A programmer in doubt about how fast to run a particular tool should consult some form of cutting conditions reference in order to come up with the recommended cutting conditions. There are numerous reference manuals as well as technical information supplied by tooling manufacturers for this purpose.

Most suppliers of this cutting-condition-related information have developed software that makes researching cutting conditions easier. With this software, the CNC user can attain the recommended cutting conditions for a given operation much faster than by looking in reference manuals.

2

Machining-Center
Accessory Devices

To this point we have been exclusively discussing devices that help with the preparation, verification, and transfer of CNC programs. Nothing at all has been discussed about programming techniques. This chapter begins a presentation of specific information related to accessory devices that are used directly with CNC equipment. In this chapter we discuss machining-center accessories. The next chapter presents turning-center accessories.

As we present each device, we will first discuss its function and application. This will give you a good idea of what the device is intended to do. If the device is available in several variations, we will present the various possibilities. If the device requires special programming commands, they will be presented as well. When applicable, an example program will also be shown.

Generally speaking, while there is no specific order to the way presentations are made in this chapter, they will move from simple and commonly used devices to those which are more complex and less commonly used. This method should make this chapter a little easier to follow.

Automatic Tool Changers

While there are still a certain number of CNC milling machines sold without automatic tool changers (ATC), most full-blown CNC machines come with them. As with many CNC accessories, they vary dramatically in makeup and programming from one manufacturer to the next. Here we will show the features and programming commands of the most popular types of ATC devices.

The obvious advantage of automatic tool changers on machining centers is that they eliminate the operator intervention involved with tool changing. In fact, the automatic tool changer is considered to be one of the largest single contributors to unmanned CNC machine-tool operation. Without the need for a person to change tools, the machine will be able to run a totally automatic cycle. The operator's only responsibility is to load and check workpieces and activate the machining cycle. This, of course, frees the operator to do other things while the CNC machine center is running production.

Each machine-tool builder will determine what it considers to be the best combination of attributes for the machine's purpose. Here are some automatic tool changer attributes each builder determines:

Tool capacity (number of tools)

Tool weight

Tool change time

Ease of use

Durability

Cost

As you might expect, the best automatic tool-changing systems are easy to program and utilize, are fast and reliable, and hold a large number of tools. Unfortunately, they are also the most costly to manufacture. In order to keep the overall machine cost down, certain compromises may be made by the machine-tool builder related to the above-mentioned desirable attributes.

Almost all forms of automatic tool changer systems require that one or more axes (X, Y, Z) be perfectly positioned in order for a tool change to occur. Most vertical machining centers, for example, require that only the Z axis be positioned at its reference point (close to the Z plus overtravel limit) before a tool change can occur. Many horizontal machining centers require that both the Y and Z axes be positioned to their reference points before the tool change.

For some machines, this motion is done automatically, as part of the tool-changing command itself. Other machines require that the programmer specifically make this positioning movement for each tool change within the program, prior to the actual tool-change command itself.

Automatic tool changer types

Generally speaking, automatic tool changers for machining centers fall into three basic categories: single-arm tool transfer systems, double-arm transfer systems, and turret-style transfer systems. Though there may be variations within each style, here we present the most basic configurations.

Single-arm tool transfer systems. This automatic tool-changer system uses a pick-and-place motion to change tools. Since the single arm can do only one thing at a time, each tool must be handled twice by the arm, once to place it in the spindle and once to put it away. Here is a basic sequence of what happens during a tool change with the single-arm tool changing system.

First, the tool that is in the spindle is removed by the single arm and placed into its original magazine position. Second, the tool magazine then rotates around until the next tool station is ready. Third, the single arm places the new tool into the spindle. Though this form of automatic tool changer tends to be relatively inexpensive to manufacture, it also requires somewhat time-consuming tool changes. Since only one arm is involved in the tool-changing motion, only one thing can happen at a time. Cycle time will suffer, especially while the magazine is rotating around to the next tool station. If the operator loads tools efficiently, placing tools that run in sequence in close proximity to each other, this wasted cycle time can be minimized. But if no special consideration is given to tool placement in the magazine, a great deal of production time will be wasted during tool changes.

Programming for single-arm tool changers. Programming for a single-arm automatic tool changer is usually quite simple. Most require that one or more axes of the machine be positioned to the ATC position *before* the actual tool change can be commanded. Most vertical machining centers, for example, require that the Z axis be positioned close to its plus overtravel limit. One common CNC control uses a G28 reference return command for this purpose.

Other machining centers incorporate this return to ATC position in the actual tool-change command itself. For these machines, the programmer need not make a special command just for the purpose of sending the machine to its reference position.

With most current single-arm automatic tool changers, only a T word is required to make a tool change. The T word simply tells the control which tool is to be placed into the spindle. The ATC system will automatically replace the tool currently in the spindle back in its original magazine position. Then the magazine will rotate the tool station specified by the T word. (Most will rotate in the shortest direction to the next tool.) Finally the tool in that magazine station will be placed into the spindle. All of this is done by the T word. As stated, on some machines, the T word even sends the machine to its ATC position in the required axes.

An example command for this type of system is

N005 T01

In this case, the tool currently in the spindle will be placed in its original tool position, the magazine will rotate to station 1, and tool 1 will then be placed in the spindle.

Most single-arm tool changers allow random access to tools in the magazine, meaning any tool from any tool station can be loaded into the spindle at any time. The tool changer need *not* follow a sequential tool-changing order. However, if tools are not called in sequential order, tool-change time for a single-arm automatic tool changer will be variable. For example, it will take longer to change from tool 1 to tool 5 than it takes to change from tool 1 to tool 2. This increase in tool-change time is a result of the amount of rotation the magazine must make during the tool-change cycle.

Many single-arm tool transfer systems require that the tool in the spindle (and in each magazine position) be properly aligned before a tool change can occur. This alignment allows a key on the tool-change arm to line up with a keyway on the tool holder during the tool change. If this is the case, the spindle must be rotated to a precise angular position before a tool change can occur. For machines that require this, the precise angular spindle position is called the *spindle orient position.*

In almost all cases, the T word itself will cause this spindle orient rotation to occur as part of the normal tool-change function; that is, the T word can still be used to command the entire tool change. Though this is the case, some time can be saved if the tool in the spindle is commanded to rotate to the orient position during a tool's return to the tool-change position.

For most machines that require the spindle orientation as part of the tool change, a separate command can be given to make the spindle orientation rotation prior to the T word. An M19 is usually the miscellaneous command that causes the spindle orient. If included within the command for each tool's return to the ATC position for every time a tool change is made, the M19 can save a great deal of time over the machine's usage period. Here is an example command that stresses the use of the M19:

N055 G91 G28 Z0 M19 (Return to ATC position, orient the spindle on the way)

N060 T02 (Put tool 2 in the spindle)

In line N055, the machine is being commanded to return to its ATC position in the Z axis. The M19 included within this command tells the machine to begin rotating the spindle to its orient position as soon as the Z axis starts moving. By the completion of the Z-axis motion, the spindle will be at its orient position, ready for the tool change to occur.

The amount of time that can be saved per tool change by incorporating this technique will vary from one machine to the next, and averages from 1 to 3 seconds per tool change. While this may not sound like much for one tool change, consider how many tool changes a machining center will make over the course of its usage. Say for example, your particular machining center will save 2 seconds per tool change if you incorporate the M19 technique. Say the machining center is dedicated to a job that includes 10 tools, and that 50 workpieces per day are machined. For a 50-week year made up of 5-day work weeks, your company will save over 34 hours of production time per year (2 seconds for each tool times 10 per workpiece times 50 workpieces times 5 days per week times 50 weeks) by this simple technique! The more tool changes you make, the more your company will save. If your machine requires the spindle orient for tool change, we *strongly* recommend that you take advantage of the M19 technique.

Double-arm transfer systems.   This automatic tool-changing system exchanges the tool that is currently in the spindle with the next tool that is currently in the magazine in one fluid motion. This action eliminates the need for the magazine to rotate during the tool change and minimizes the time required for the tool change to occur. It also makes tool-changing time consistent from one tool change to the next, since the magazine need not rotate a variable number of stations during the tool-change cycle. Figure 2.1 shows an example of a double-arm tool-changing system.

Generally speaking, this form of ATC must be a little more intelligent than the single-arm automatic tool changer. The CNC control has to keep track of the tool station numbers automatically. For example, say that tool 1 is currently in the spindle. At this point, say that a change to tool 5 in tool station 5 is commanded. During the tool change, tool 1 will be placed in the tool station originally occupied by tool 5. In this case, the control will have to rename the tool station that was previously tool station 5 as tool station 1.

Though some (older) machining centers have a holding station in the automatic tool-changer system and handle the problem mechanically, most current ATC systems simply utilize computer logic to renumber tool stations as tools are changed.

The major time savings advantage of the double-arm automatic tool-changing system over the single-arm type is that the magazine can begin rotating to the next tool station *before* the tool change is

Figure 2.1 Double-arm-style automatic tool changer. (*Courtesy Okuma & Howa USA.*)

commanded. As the tool in the spindle is machining the workpiece, the tool magazine will rotate, bringing the next tool into the ready position. When the tool change is made, the double arm will rotate, engaging one tool on each side of the double arm. The double arm then pushes the tools out of the spindle and ready position and rotates about 180 degrees. Finally, the double arm pulls the tools into the spindle and ready position and swings out of the way.

There are usually two programming commands required for this type of ATC. A T word tells the control to rotate the tool magazine, bringing the designated tool into the ready position. The T word by itself does *not* actually make the tool change. It just gets the next tool ready. An M06 command tells the control to actually make the tool change. Whatever tool is in the waiting position will be exchanged with the tool in the spindle when an M06 is executed.

A simple example for this form of ATC is

O0001 (Program number)
N005 T01 M06 (Place tool 1 in spindle)
N010 G90 S500 M03 T02 [Select absolute mode, turn spindle on clockwise (CW) at 500 RPM, get tool 2 ready]
N015 G00 X0 Y0 (Rapid to first XY position)
N020 G43 H01 Z.1 (Instate tool length compensation, rapid to first Z position)
N025 G01 Z-.75 F5. (Drill hole)
N030 G00 Z.1 (Rapid out of hole)
N035 G91 G28 Z0 M19 (Rapid to tool change position in Z, orient spindle on the way)
N040 M01 (Optional stop)
N045 T02 M06 (Place tool 2 in spindle)

N050 G90 S600 M03 T03 (Select absolute mode, turn spindle on CW at 600 RPM, get tool 3 ready)

In line N005, since tool 1 is the first tool to be used in the program, it is being placed in the spindle. In line N010, as tool 1 begins, note that the word T02 will rotate the magazine, getting tool 2 ready as tool 1 goes to work on the workpiece. At the completion of use of tool 1, line N045 places tool 2 in the spindle. In line N050, as tool 2 begins, tool 3 is told to rotate to the ready position. This technique is repeated for the balance of the program. As the last tool begins, tool 1 will be rotated to the waiting position (since it is the program's first tool).

As stated, the M06 command will place whatever tool is in the ready position into the spindle. So you may be questioning why, in line N045, the T02 word is repeated. Truly, tool 2 has been commanded in line N010, and should still be in position when line N045 is executed. Though the T word in line N045 is not absolutely mandatory, we recommend repeating the desired T word within every tool-change command (M06) for two reasons.

First, it is possible that as the machine is machining the workpiece, the operator may wish to replace worn tools in the magazine. If the operator manually rotates the magazine during the machine's automatic operation, the magazine may be left out of position. If the T word is not repeated within every tool-change command, it is possible that the wrong tool may be placed into the spindle, causing disastrous results.

Second, repeating the T word within every tool-change command makes it easier for the operator to pick up in the middle of the program. If the operator wishes to rerun a tool, having the T word within each tool change command will make the task much easier. For example, in the preceding program, if the operator wishes to rerun only tool 2, he or she can simply begin the cycle at line N045. If the tool station number was not in this command, beginning from tool 2 would not be so easy. In this case, the operator would have to manually load tool 2 into the spindle *before* the tool could be rerun.

As with the single-arm transfer system, most double-arm automatic tool changers require that the tool in the spindle be properly oriented rotationally before a tool change can occur. Notice in line N035 the M19 is getting the spindle to orient on the tool's return to tool-change position.

Turret-style tool transfer systems. This form of automatic tool changer stores a limited number of tools (usually only six or eight). Each tool is in a station that resembles the spindle itself. That is, many spindle functions are built into each turret station and each turret station is actually designed to act as the spindle itself. After the turret rotates and brings the tool to the spindle position, the tool will be ready to go to work on the workpiece.

While this turret design allows for quite a fast tool change, one dramatic limitation of this system is the number of tools that can be used by each program. As stated, usually only six or eight tools can be accommodated by this style of ATC.

Programming this style of tool changer is quite simple. Usually only a T word is required to specify which turret station is to be used. For example, the command:

N005 T01

will rotate the turret to bring tool station 1 to the machining position.

Pallet Changers

For complicated setups, it can conceivably take as long to load a workpiece into the setup as it does to run the entire machining cycle. This is especially true for multiple workpiece setups. For many machining centers, this is wasted production time, since the machine will sit idle while the operator is loading workpieces.

Pallet changers can dramatically reduce this nonproductive workpiece loading time. They allow the operator to safely load one workpiece (or a series of workpieces) into one setup while another is being machined. At the completion of the machining cycle, the entire setup in the loading position is exchanged with the setup in the machining position. Workpiece loading time can be kept consistent and equal to the amount of time it takes the pallet changer to make the exchange (unless it takes longer to load workpieces into the setup than it does to complete the machining cycle).

Many of the same desirable attributes for automatic tool changers apply to pallet changers. They include:

Pallet-changing time

Load capacity in weight

Durability

Pallet dimensional capacity

Accuracy and repeatability

And as with automatic tool changers, the best pallet changers are fast, accurate, and durable.

One of the primary concerns with any pallet changer is repeatability. The repeatability of the pallet changer is often directly related to the best possible workpiece accuracy, especially if an operation has been performed on the workpiece prior to the current machining-center operation. Of course the pallet changer must be capable of precisely locating each pallet into the work position over and over again. If there is any deviation from one pallet change to the next, the deviation will almost always have a direct impact on workpiece accuracies.

Pallet changers are most commonly used with horizontal machining centers. Since most horizontal machining-center applications require rather complicated setups and lengthy machining cycles, they are perfect candidates for pallet changers.

Though good applications are not quite as common, pallet changers can be equipped on most vertical machining centers as well. When attached to vertical machining centers, the placement of the pallet changer is usually to the right of or directly in front of the machine.

Pallet changer types

There are three different styles of pallet changers we will examine. They include manual, rotary, and shuttle types.

Manual pallet changers. Manual pallet changers are most commonly found as aftermarket devices attached to vertical machining centers. By *aftermarket,* we mean they are purchased by the CNC machine end user separately from the machine tool itself. Since they are usually quite simple in design, manual pallet changers are commonly installed by the machine's user.

The actual pallet exchange is accomplished manually by the operator. When the pallet change is required, the operator will unlock the pallet in the working position by releasing one or more clamps. Then the operator will manually push this pallet along a series of rollers until it is engaged with the pallet-changing device. The pallet changer will then be rotated, swinging the workpieces to be machined toward the work area. Finally, the operator will push the pallet into the work area and tighten the clamps.

While all of this may sound a little cumbersome, most manual pallet changers are quite easy to work with, and the time required to complete the entire pallet change is usually under 1 minute.

Keep in mind that in order for a manual pallet changer to be the most efficient method of workpiece loading, the operator must be able to complete the pallet change in a shorter time than it takes to actually load a workpiece into the setup. If for example, the operator must simply clamp one workpiece in a vise, applying a manual pallet changer may actually add time to the workpiece loading operation.

For jobs that do not require the use of the manual pallet changer, the pallet-changer mechanism must be moved out of the way to allow the operator to easily work around the table area. Though some manual pallet changers must be permanently mounted to the machine tool itself, most are mounted on sliding surfaces to allow easy movement of the entire pallet-changing device.

Rotary-style automatic pallet changers. This form of automatic pallet changer is usually mounted to the right side of or in front of the machine tool. It incorporates a rotary motion with which to swing the loaded workpiece setup into the work position. This rotary motion resembles a double-arm automatic tool changer. Depending on size and durability, this form of automatic pallet changer will completely perform the pallet change in about 30 to 60 seconds. Figure 2.2 shows a rotary-style automatic pallet changer.

Most machine-tool builders who incorporate this form of automatic pallet changer utilize a simple M code with which to make the pallet change. The most common M code used for pallet change is M60. Once commanded, the M60 performs the entire pallet exchange.

As with automatic tool changers, one or more of the machine's axes must be properly positioned *before* a pallet change can take place. Most vertical machining centers, for example, require that the X and

Figure 2.2 Rotary-style automatic pallet changer. (*Courtesy Okuma & Howa USA.*)

Y axes be positioned very close to their positive limits at their reference return position prior to a pallet change. Most horizontal machining centers require that the X and Z axes be so positioned before a pallet change can occur.

For safety reasons, all automatic pallet changers will incorporate some form of safety button or switch that will keep the pallet change from occurring until the operator has totally completed the loading of the workpieces. In most cases, the operator must push the button or turn on the switch each time he or she is finished with the workpiece-loading procedure. If the machine completes its cycle and the pallet-change command is read, the control will look to the position of this standby button or switch. If it has not been activated, the machine will wait to perform the pallet change until the operator activates the button or switch.

Keep in mind that the operator can activate the standby button or switch at any time during the machining cycle. If the operator completes the setup quickly, the standby function can be immediately activated. In this case, the pallet change will occur as soon as it is commanded. Also note that many automatic pallet changers allow the operator to have a change of heart. If for example, the operator notices a problem with the setup and wishes to cancel the standby mode, the standby switch can simply be turned off.

Shuttle-style automatic pallet changers. This form of automatic pallet changer is commonly mounted in front of the machine. The machine table (usually the X axis) moves into a position that will allow the pallet to be removed from the work area. The pallet in the work area is then pushed onto the shuttle device. The table (X axis) then moves to the other side, where the pallet in the loading area is pushed onto the machine table. The shuttle system then retracts and the shuttle transfer is complete. For automatic pallet changers of equal capacity, this form of pallet changer is usually slightly faster than the rotary style. Normally an automatic pallet shuttle system can complete its cycle in 10 to 25 seconds. Figure 2.3 shows a shuttle-style automatic pallet changer.

The M60 word is still a very popular command used to make the pallet change. However, several machine-tool builders utilize two subprograms to command the pallet change. These pallet-changing subprograms are permanently stored within the CNC control's memory.

For example, one popular machine-tool builder uses subprogram O9000 to load the left pallet (we'll call it pallet A) and subprogram O9001 to load the right pallet (pallet B). To load pallet A, the programmer simply executes subprogram O9000. To load pallet B, subprogram O9001 is executed.

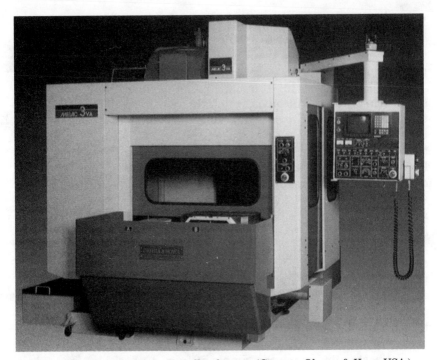

Figure 2.3 Shuttle-style automatic pallet changer. (*Courtesy Okuma & Howa USA.*)

Here is an example main program that runs two different work-pieces. Subprogram O0001 is the program that machines the work-piece on pallet A, while subprogram O0002 is the program which machines the workpiece on pallet B. For this particular control, the M98 command tells the control to execute the subprogram specified by the P word. We'll say pallet B is currently in the work position and pallet A is in the loading position.

```
O100 (Main Program)
N005 M98 P9000 (Load pallet A)
N010 M98 P0001 (Run workpiece on pallet A)
N015 M98 P9001 (Load pallet B)
N020 M98 P0002 (Run workpiece on pallet B)
N025 M99 (Return to the beginning of this program and continue)
```

In line N005, pallet A is loaded. In line N015, pallet B is loaded. Again, notice that even the machining programs (O0001 and O0002) are subprograms. This makes it very easy to keep the machining program needed for pallet A separate from the machining program needed for pallet B. This is especially helpful when there is no relationship between the two setups being machined. Two completely different workpieces could be machined without the need to merge the programs required for each workpiece into one program.

To ensure a completely automatic cycle, notice that this main program ends with an M99. For this particular control, when a main program ends in this manner, the control will simply return to the beginning of the program and continue without stopping. When the control reaches line N025, and as long as the operator has activated the standby button (discussed during the presentation of rotary-style pallet changers), it will immediately flow to line N005 and continue automatic operation by changing pallets. The operator will *not* have to press the cycle start button at the completion of this program.

If two identical workpieces (or setups) must be run (one on each pallet), the same program can be called for in lines N010 and N020. This program structure (utilizing subprograms) still makes for very efficient and effective control of the pallet-changer device while allowing two different workpieces to be machined.

Multiple-pallet changers. Another, more elaborate form of automatic pallet-changing system allows more than two pallet stations. Most allow from 6 to 10 pallet stations, and some even allow up to 30. This type of pallet changer is most commonly used in a flexible manufacturing system (FMS), which allows specific and timely instructions to be given for the production of workpieces. Figure 2.4 shows a multipallet automatic pallet changer.

Generally speaking, these pallet changers are managed by a computer related in one manner or another to the company's production control department. This computer places a priority on each workpiece in the system and automatically selects the sequential order by which workpieces are to be machined.

With this kind of automatic pallet-changing system, workpiece loading is usually done off line. The operator will not actually load a workpiece in one of the pallets currently in the system. A special loading station will be used for this purpose. Only after the workpiece is loaded will the pallet be engaged into the automatic pallet-changing system and logged into the pallet conveyor.

The ultimate goal with this type of system is totally unmanned operation. While problems may still exist with tool life, and cutting conditions may have to be slowed to ensure a margin of safety, many companies utilize multiple-pallet changers even when there is no operator present. They leave the machine unattended, in many cases even running through the night. Some machines that have this kind of pallet-changing system even have a toggle switch labeled *night work,* which if activated, will automatically turn the machine off after the last workpiece in the system is machined.

Figure 2.4 Multiple-pallet automatic pallet changer. (*Courtesy Okuma & Howa USA.*)

In order to be safe and feasible to use, this type of system almost always requires more than a simple multipallet automatic pallet changer. They require other accessory devices to ensure the proper usage of the FMS system. These sophisticated machines will usually have tool-life monitoring systems to allow more flexibility with cutting tools. They may also have probing systems to perform functions the operator will not be available to perform. Some have elaborate tool-breakage detection systems that will automatically stop production in case of a broken tool. Some are even tied into a phone line and will automatically place a phone call to the operator or setup person (at home) in case the machine shuts down for any reason! These sophisticated devices are discussed later in this chapter.

Rotary Devices

Rotary devices are applied to many forms of CNC equipment. The basic machine style has a great deal to do with exactly how the rotary device is applied. Keep in mind that the presentations made in this chapter are strictly related to how rotary devices are applied to machining centers. In the next chapter, we will discuss how they apply to tuning centers.

When a rotary device is applied to a vertical machining center, it is simply clamped to the tabletop. This means it can be easily removed for those jobs that do not require the rotary device. Many times the rotary device for a vertical machining center itself is not even purchased from the machine-tool manufacturer; rather it may be purchased directly from the manufacturer of the rotary device.

On the other hand, when a rotary device is applied to a horizontal machining center, it is normally an integral part of the machine tool itself, manufactured and assembled by the machine-tool builder. It is located within the table mechanisms of the machine and cannot be removed.

Types of rotary devices

Generally speaking, rotary devices fall into two distinctly different categories, indexers and rotary tables. Likewise, there are two basic purposes for utilizing rotary devices. The primary purpose of any rotary device is to expose different surfaces of the workpiece to the spindle during the machining cycle. When used for this purpose, the user can dramatically reduce the number of setups needed to machine complex workpieces. For example, if a workpiece must be machined in two or more attitudes, the rotary device can be used to rotate the workpiece, exposing the different surfaces to be machined. Both indexers *and* rotary axes easily allow this primary purpose for rotary devices to be accomplished.

Machining several surfaces of the workpiece in one setup makes it easier to hold accuracy from one surface of the workpiece to the next. If multiple setups must be made, workpiece accuracy may suffer since it is difficult to perfectly locate the part in each setup. Imperfections will accumulate from one setup to the next.

An indexer or rotary table eliminates this accuracy problem since the workpiece is not being removed from the setup for the purpose of machining each surface. For this reason, accuracy from surface to surface will dramatically improve.

Another, somewhat less popular, purpose for rotary devices is to allow machining of the workpiece during rotation. *Only* a true rotary axis (rotary table) is capable of this function. That is, an indexer *cannot* allow machining during the indexing of a workpiece. This kind of operation is often necessary for machining cams, where an end mill is used to machine the workpiece during a rotary axis motion. Possibly, a combination of rotary axis motion and another (linear) axis motion may even be required during a machining command.

Indexers. An indexer is a rotary device that allows the workpiece to be quickly rotated a certain angular amount. The rotation speed is usually very fast and the programmer will have no control of the rotation rate. The indexer will simply rotate as fast as it can. This, as stated, means that it is impossible to be machining the workpiece while the indexer is rotating. Though there are exceptions, most indexers can rotate only in one direction, therefore the rotation to the desired angular position may not always be the in most efficient direction.

There are many kinds of indexers, but the most basic designation for an indexer has to do with the smallest angular increment the indexer is capable of rotating. This smallest angular increment will also determine the method by which the indexer is programmed. Common indexers include those with minimum increments of 90°, 45°, 5°, and 1°.

As you might imagine, the smaller the angular rotation possible, the more flexible the indexer. A 1° indexer has 360 possible rotational positions, while a 5° indexer has only 72. Of course, a 90° indexer has but four index positions.

Unfortunately, flexibility requires that a price be paid in rigidity. Generally speaking, the more index positions possible, the less rigid the indexer. This is simply due to the mechanical linkages required. The more index positions allowed, the more mechanical linkages required within the indexer, and the weaker the indexer will be.

The methods of programming indexers. The 90° and 45° indexers are usually activated by a single M code. The M code number is determined by the machine-tool builder or by the company supplying the indexer. It can be found in the list of M codes that comes with the machine tool or indexer. Of course, the programmer *must* know the M code number required to command indexes.

With this kind of indexer, when the programmer wants an index to occur, the M code is commanded and *one* index of the predetermined angle will occur. This form of indexer programming can be somewhat cumbersome to work with. Say, for example, you have a 45° indexer and you want to index 180°. Say an M13 is the command that causes a 45° index. In this case, four successive M13 commands must be programmed (180 ÷ 45 = 4). Keep in mind that most CNC controls allow only one M code per command.

The 5° indexers vary further. Some builders give you a series of M codes for their 5° indexer, based on the most common angles of index. For example, M71 may be used to specify a 5° index, M72 a 15° index, M73 a 45° index, and M74 a 90° index. This would minimize the number of M codes the programmer would have to string together for an odd index angle.

Other 5° indexers force the user to mechanically set the angle of index desired per rotation. This means the user will have to manually set the indexer to the smallest angular departure required by the program as part of the setup. This form of indexer is also quite cumbersome to work with, since several M codes may have to be run together in order to rotate to an odd index angle.

The 1° indexers are usually the easiest indexers to work with. Most 1° indexers are programmed with a special letter address word. Though the word may vary from one control manufacturer to the next, most use the letter address B to command indexing. With the B word, a programmer can easily specify the exact angle of index desired. If the programmer wants a 27° index, the command B27 is specified.

Many 1° indexers even allow the programmer to specify the direction of rotation. Two M codes are used for this purpose. For example, an M12 may specify that rotation is in a clockwise direction, while M13 specifies counterclockwise rotation.

Since the mechanical linkages required within a 1° indexer tend to reduce the rigidity of the indexer itself, many indexer manufactures allow the indexer to be clamped once the index angle is reached. For this purpose, two more M codes may be involved, one to clamp the indexer and the other to unclamp. This feature dramatically improves the rigidity of the indexer.

As you have seen, there are several possibilities for how any one CNC machining center's indexer may be programmed. When working with any new machine, you must check in the machine-tool builder's manual or indexer manual to attain the method by which your particular indexer is programmed. This presentation should have prepared you for most of the possibilities.

Keep in mind that knowing the actual programming commands needed to cause the indexer to rotate is only part of the entire rotary device programming problem. As you know, each machining-center program will utilize a program zero point from which all dimensions in the program are taken (assuming you program in the absolute mode). After an indexer rotates, it is likely that the program zero point used for the surface prior to rotation will not make the best program zero point for the surface to be machined after the rotation. After we present the basic commands related to rotary tables, we will address how this program zero problem can be handled.

Rotary tables.   As with indexers, a rotary table can be applied to both vertical and horizontal machining centers. Also as with an indexer, it is usually mounted on top of the table of a vertical machining center, and mounted within the machine as part of the table mechanism of a horizontal machining center.

A rotary table is much more flexible than an indexer for several reasons. First, the method by which rotation is commanded is much more precise. The programmer can control almost perfectly the rotation angle desired, and is allowed to command rotary departures to within 0.001°. For this reason, a rotary table can be thought of as an indexer having 360,000 positions!

Second, the rotary table motion command allows a decimal point to specify portions of a degree (just like the X, Y, and Z axes). This makes it very easy to specify angular positions to an accuracy of less than 1°. For example, if an ending angle of 45.75° is desired, and if the letter address C is used to command rotary axis motion, the word C45.75 will be commanded.

Third, the direction of rotation (clockwise or counterclockwise) is relatively easy to command. This means the programmer can easily command that the rotary device take the shortest rotation direction to the desired angular position.

Fourth, the programmer can control the rotation rate (feed rate) at which the rotary table turns. This is what gives the rotary table its ability to be machining and rotating at the same time.

Fifth, most rotary tables can usually be automatically clamped once a rotation is completed. This improves their rigidity for heavy machining operations.

How the rotary axis is named. The rotary motion of a rotary table is actually a true axis of the machine. This means that rotary axis departures are commanded with a letter address, just like the X, Y, and Z axes. The designated letter address for the rotary axis depends on the machine-tool builder as well as how the rotary axis is positioned on the machine tool.

The standard naming conventions for the rotary axis are as follows: If the rotary axis is parallel to the X axis, it should be named the A axis. If parallel to the Y axis, the rotary axis should be named the B axis. If parallel to the Z axis, it should be named the C axis.

By the parallel axis, we mean the axis which is in line with the center of rotation. Imagine a long piece of round stock concentric with the rotary axis. To determine the parallel axis, ask yourself which linear axis the length of the round stock is running along. This will be the parallel axis.

On horizontal machining centers, for example, the rotary device is rotating the table itself. In this case, a long piece of round stock concentric to the rotary axis would be running along the Y axis (vertically), so the axis parallel to the rotary axis is the Y axis. In this case, the designation for the rotary axis will *always* be the B axis (no horizontal machining center manufacturer we know of strays from this standard). For a vertical machining center, if the rotary axis is mounted horizontally, parallel to the X axis (as it normally is), the rotary axis should be called the *A* axis. If it is mounted vertically (parallel to the *Z* axis), it should be named the *C* axis.

Unfortunately, when it comes to vertical machining centers, some machine-tool builders stray from this standard of how the rotary axis should be named, and you must be prepared for variations. For example, some call the rotary axis on the vertical machining center the *B* axis. Still others call it the *C* axis. Also, many rotary tables are designed to be mounted in one of two positions, one parallel to the *Z* axis, and the other parallel to the *X* axis. The rotary table itself can be flipped 90° to allow this. Most machine-tool builders do not change the name designation for the rotary axis simply because it has been tilted 90° on the machine table.

Whether the machine-tool builder followed the standard naming conventions for a machine's rotary axis (*A, B,* or *C*) is not nearly as important as the programmer's knowing what the axis name is for a particular machining center. It really doesn't matter if the machine-tool builder sticks to the standard as long as the programmer knows how the machine's rotary axis is designated.

For the purpose of this text, we will call the rotary axis the *B* axis for horizontal machining centers and the *C* axis for vertical machining centers. If you understand our presentations, you will be able to easily apply what you have learned to your particular machine, no matter how the rotary axis is named. Also note that, because of this axis-naming confusion, some machine-tool builders designate the rotary axis (on the machine's control panel) as "fourth" and do not put the letter address related to the rotary axis on the machine panel at all.

How to program rotary-axis departure.   All rotary commands are given in degrees of rotation. For example, if a programmer wants to designate a rotary axis position of 45° and if the axis letter address for the rotary axis is C, here is the way it would be specified:

N050 C45.

Notice that a decimal point is allowed within the rotary axis designation. It is important to know that the rotary axis can only be programmed to three places after the decimal point (*not* four, unlike *X, Y,* and *Z*). Also, if the decimal point is accidentally left out of the rotary-axis command, the control will use the fixed format for the word. For example, say the programmer intended to command a 45° departure but omitted the decimal point. The command

N055 C45

would *not* be taken as 45°. Instead, it would be taken as 0.045°. The control would automatically place the decimal point (incorrectly in this case) three places to the left of the rightmost digit. If the programmer wishes to program without a decimal point (as programmers of some computer-aided manufacturing systems do), 45° could be commanded as

    N055 C45000

Note the three zeros to the right of the value.

Here is another important point about angular departure, with or without the decimal point. If you are designating an angular departure that includes a value of less than 1°, you *must* make the designation in *decimal portions of a degree*. Many workpiece drawings designate portions of a degree in minutes and seconds. Minutes and seconds *must* be converted to decimal format in order to correctly command rotary departure for a CNC machine. Here is the formula to convert minutes and seconds to decimal format:

Decimal format degrees = degrees + minutes ÷ 60 + seconds ÷ 3600

Say, for example, the value of 13°, 27 minutes, 37 seconds is on a workpiece drawing. To input this angular value into a rotary-axis command in a CNC program, it would first have to be converted to decimal format.

To make the conversion, first 27 minutes is divided by 60. The result is 0.45°. Then 37 seconds is divided by 3600. The result is 0.010°. Then 13° is added to 0.45 and 0.010. The result is 13.46°. And, if the rotary axis is designated with the letter address C, the motion would be specified as

    N045 C13.46

It can be a little frustrating if drawings are dimensioned with minutes and seconds for the angular dimensions. But decimal format is the *only* way a CNC control will understand angular values.

Comparison to other axes. Just about everything you know about the X, Y, and Z axes will still apply to a rotary axis. That is, the methods by which normal motion is commanded in X, Y, and Z can still be used to control the rotary axis. Here is a list of the things we will discuss about the rotary axis that should be somewhat familiar if you understand motion in X, Y, and Z:

1. Reference position

2. Designation of program zero

3. Absolute mode (G90) and incremental mode (G91)

4. Rapid (G00) and straight line cutting (G01)

5. Usage of canned cycles and other special programming features

Truly, rotary-axis motion is treated as any other axis motion. The next few discussions will draw on what is known about the X, Y, and Z axes to help explain the rotary axis.

Reference position.   Just like X, Y, and Z, a rotary axis will have a reference position. Part of powering up the machine will be to return the rotary axis to its reference position just as is done with the X, Y, and Z axes. The accuracy of the reference position for the rotary axis is just as important as it is for all other axes.

To command the rotary axis to go to the reference position in the program is very similar to commanding the other axes to go to zero return position. If the rotary axis is designated with a C letter address, here is a command used by one popular control manufacturer that will send the machine to its reference position in *all* axes. This particular control uses a G28 to command a return to the reference position. Any axis included within the G28 command will end up at the reference point.

    N055 G91 G28 X0 Y0 Z0 C0

In this case all four axes would move to their reference point position at the same time. If only the C axis is to be sent to its reference position, here is the slightly modified command:

    N055 G91 G28 C0

As with X, Y, and Z, the reference position for the rotary axis will make an excellent point of reference for a program. The angular program zero point in the rotary axis can easily be taken from its reference position.

This next point may be a little difficult to visualize. With X, Y, and Z, the motion is along a linear path. There is a definite limit as to how far a linear axis can move in either direction. In fact, any linear axis will overtravel, generating an alarm, when a limit is reached.

On the other hand, a rotary axis has no such physical limit. It can continue rotating countless revolutions in either direction (clockwise or counterclockwise) without ever coming to a limit, meaning a rotary axis will never overtravel as a linear axis would. This actually presents a small problem when you are making the command to send the rotary axis to its reference position.

When commanding that an axis be sent to its reference position, most CNC machines require this motion to be specified in the plus direction. As the reference command is given, the control will begin searching for the reference return position in the plus direction. For X, Y, and Z, since they are linear axes, this presents no problem, since the reference position is normally close to the machine's plus overtravel limits. It will always search in the correct direction (plus).

Depending on the rotary-axis position at the time a command is given to send the rotary axis to its reference position, the rotary axis may choose a somewhat inefficient direction of motion. For example, if the rotary axis is resting at a position of just slightly minus of the reference position when the reference return command is given, the rotary axis will make only a small (and efficient) rotary motion to return to its reference position. On the other hand, if the rotary axis is resting just slightly plus of its reference position when the reference return command is given, it will take the *long* way (plus), rotating almost one full revolution to reach its reference position. While the rotary axis will eventually reach its reference position, the motion will not be very efficient.

For this reason, and especially when commanded in a program, the programmer must always try to begin the rotary axis return to its reference position from an angular position just slightly minus of the reference position. This can sometimes be difficult to visualize while programming. However, if no concern is given to this potential problem, cycle time will suffer during inefficient table rotation to the machine's reference point.

Designation of program zero. If programming the rotary axis in the absolute mode (G90), the programmer must designate a program zero location for the rotary axis. As with X, Y, and Z, the control must know a point of reference (the program zero point) from which all coordinate values in the program are to be taken. Also as with X, Y, and Z, the programmer must include the rotary axis (A, B, or C) as part of the designation of program zero.

The two most popular ways to designate program zero are from within a program (with a G92 command) and by some form of fixture offset. Either way, if the machine has a rotary axis, the program zero designation method will allow the designation of program zero within the rotary axis.

How the rotary axis program zero point is designated is always the same as for X, Y, and Z. When using G92 to designate program zero from within the program, the angular distance *from* the rotary axis program zero point *to* the rotary table's reference position is the value to be included within the G92 command (assuming the program begins with the machine resting at its reference point in all axes). For example, say you intend to make the program zero point for the rotary axis the same angular location as the reference return position. Here is an example G92 command that would accomplish this.

N005 G92 X15.9272 Y12.4733 Z15.3933 C0

For the C axis, this command is simply telling the control that the angular distance from program zero to the reference position is nothing. From this point, *all* angular designations in the absolute mode can be taken from this program zero point in the C axis.

If using fixture offsets (G54–G59 on one popular control), the measurement will be taken *from* the reference position in the rotary axis *to* the program zero point (just like X, Y, and Z). This simply means the sign (plus or minus) must be reversed when you are using fixture offsets as compared to using a G92 command to assign program zero.

Since there are typically no angular datum surfaces on the workpiece drawing, generally speaking, there is no need to be overly picky or fussy about the location of program zero within the rotary axis. With the X, Y, and Z axes, if the program zero point is chosen wisely, many program coordinates can be taken right from the print. This is not often the case with the rotary axis. For this reason, most programmers will simply specify that the side of the rotary table facing the spindle while the rotary axis is at the reference point specified is the program zero side (as our previous example demonstrated).

As we get deeper into rotary-axis programming, you will see many instances when it may be wiser to program the rotary axis in the incremental mode instead of using the absolute mode, especially if the rotary axis is used only for indexing purposes (*not* machining and rotating at the same time). If the rotary axis is to be programmed exclusively in the incremental mode, there will be no need to even assign a program zero point for the rotary axis.

Understanding the absolute mode for the rotary table. As you know, when you work from the program zero point and input all program coordinates from that point, you are working in the absolute mode. Most beginning programmers are instructed to work exclusively in the absolute mode, since coordinates going into the program can be taken directly from the workpiece drawing. This is the easiest way for beginners to become familiar with CNC programming.

However, when it comes to the rotary axis, working exclusively in the absolute mode may not be the easiest or best way to input rotary axis departures, *especially* if the rotary axis is being used as a simple indexer. Some odd and unexpected things can happen if you do not understand how the absolute mode effects the rotary axis. Though the absolute mode truly affects the rotary axis in the same way it affects *X, Y,* and *Z,* the effects of the absolute mode on the rotary axis are harder to visualize.

With any linear axis, the direction any one motion command will generate is easy to visualize and predict. For example, on a vertical machining center, say a tool is currently resting at a position of X5.0 when the command X6.0 is given (in the absolute mode). Even beginning programmers can easily visualize that the tool will move in the plus direction and the amount of departure will be 1 in. For this machine, the plus direction means the tool will move to the right (or table to the left).

With any rotary axis, the plus direction is usually a clockwise rotation as viewed from above the rotary axis (though some rotary table manufacturers reverse plus and minus). When a rotary axis departure is given in the absolute mode, the programmer can rest assured that the designated angular position will be facing the spindle at the completion of the command. For example, if the command C45. is given in the absolute mode, the rotary motion will end with the 45° side of the rotary table facing the spindle.

However, also as with any linear axis, the direction the table rotates depends on the prior rotary axis position. If the rotary axis was previously at a smaller angular position (under C45.), the rotation will be in the clockwise (plus) direction. If the prior rotary axis position was larger than C45., the direction of rotation will be counterclockwise (minus). While this may not seem too complicated, there will be times when even an experienced programmer may be fooled and make mistakes with regard to commanding efficient rotary table rotation. Let's look at several examples that stress why.

Figure 2.5 is a drawing of a horizontal machining center as viewed from above. Keep in mind that the points made here will also apply to a rotary axis applied to a vertical machining center.

As you can see, we have depicted the rotary axis at its reference position. The designation in the fixture offset or G92 B has been set to zero. This means that the table side facing the spindle while the B axis is at its reference position is the zero side of the rotary table in the absolute mode. In the absolute mode, whenever the control reads and executes a rotary axis command of B0, the rotary table will end its rotation with this side facing the spindle.

Here's where it gets a little complicated. You *must* remember that when programming a rotary axis departure in the absolute mode, the table side commanded will end up facing the spindle at the completion of the command. As stated earlier, the direction the table rotates (clockwise or counterclockwise) depends on the position of the table prior to the command.

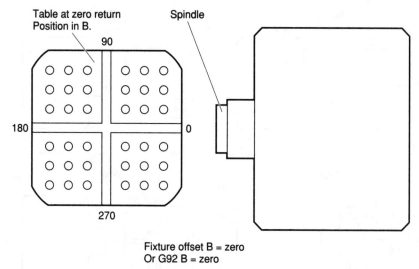

Table at zero return Position in B.

Spindle

Fixture offset B = zero
Or G92 B = zero

Figure 2.5 View of rotary table from above horizontal machining center.

Say for example, the 270° side is facing the spindle right now. In the absolute mode, if the command

    N055 G00 B0

is given to rotate at rapid to the 0° side, the control would rotate the table counterclockwise to make the motion. In essence, it would be taking the long way to get to the 0° side.

Look at the example drawings in Figure 2.6 showing the rotary table's motion in the absolute mode to get the idea. Keep in mind that these drawings apply to a rotary table in either a vertical or horizontal machining center. By viewing these examples, you may start to agree that the absolute mode may not be the best mode to work with when using the rotary axis for simple indexing purposes.

In order to handle this plus-versus-minus problem related to the rotary axis in the absolute mode, some (but not many) machine-tool builders give a special way of specifying the direction of rotation. Most that do use a series of M codes. For example, M18 may specify the neutral condition of the rotary axis. If under the influence of this command, the rotary axis will respond as described in the previous presentation (plus rotary departure being clockwise, and minus being counterclockwise). M19 might be used to specify a clockwise rotation to the rotation's end point, regardless of a plus or minus axis departure. In like manner M20 may be used to specify a counterclockwise rotation. Documentation related to these special M codes (if they are

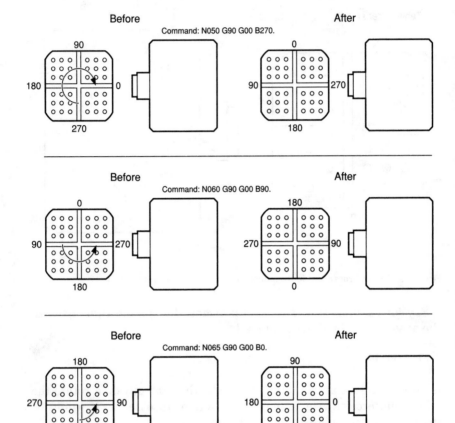

Figure 2.6 Examples of how the rotary axis is commanded in the absolute mode.

available) can be found in the machine-tool or rotary table programming manuals.

Before we end our discussion of the absolute mode, we want to make one more point. The rotary axis allows you to make commands over 360°. That is, if needed, you can make the table rotate more than once in one command. For all intents and purposes, the 360° side is the same as the 0° side of the table. The only time we recommend commanding rotation past 360° is for making cutting movements (G01) when more than one rotation of the rotary table is required.

As a beginner, and when using the absolute mode even for simple indexing purposes, you may be tempted to continue rotating the table to higher and higher rotation values. That is, as you continue rotating the table for the various tools in your program, you may be tempted to rotate well past the 360° side. If, for example, the table has to be rotated 5 times during the program for operations on all four sides of the table, your *B* axis would have accumulated to 800° (5 × 360) in the absolute mode. If, at any time during the program, you make a command to go back to the zero side, the table would unscrew all the way back to the zero side (rotating 5 times) and waste a great deal of time.

The incremental mode—best for simple indexing. The incremental mode allows the programmer to be very specific about how the rotation of the rotary axis is to occur. The direction of motion (plus or minus) as well as the precise amount of desired index can be easily specified.

As long as the rotary table is being used to simply index from one position to another, with no machining during the rotation, we recommend that the programmer temporarily switch to the incremental mode (G91) to make the rotary motion. After the rotary axis motion, the programmer should immediately switch back to the absolute mode to continue machining with *X*, *Y*, and *Z*.

In the incremental mode, plus is still clockwise and minus is still counterclockwise. But the point of reference will now be the table's current position, not the program zero position. If the programmer wants to simply rotate 90° clockwise from the current table position, this command would be given:

N050 G91 G00 B90.

If *X*, *Y*, and *Z* movements are to be programmed in the absolute mode, the programmer must not forget to switch back to the absolute mode in the next command. Of course, if the programmer wants to rotate 90° counterclockwise from the table's current position, the sign of the rotary-axis address would simply be reversed, as this command shows:

N050 G91 G00 B-90.

Look at Fig. 2.7 for more examples of how rotary-axis motions are made in the incremental mode. As you can see, this is easier and more logical than the previously described absolute method of commanding rotation.

The only possible problem with making incremental table rotations has to do with *X*, *Y*, and *Z*. There may be times when the programmer must command a table rotation at the same time a positioning motion is required in *X*, *Y*, and/or *Z*. This is commonly the case when the programmer is trying to keep the cycle time down to its bare min-

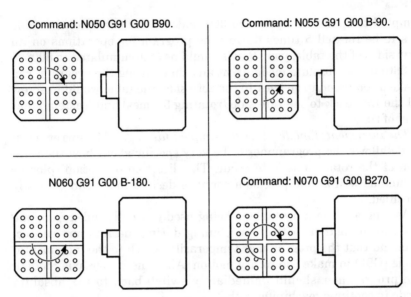

Figure 2.7 Examples of how the rotary axis is commanded in the incremental mode.

imum. If you intend to move in *X*, *Y*, and/or *Z* while rotating the rotary table, you will have to make a somewhat difficult choice. Either the rotary motion must be made in the absolute mode (making *X*, *Y*, and *Z* departures easy) or the *X*, *Y*, and/or *Z* must be made incrementally (making rotary-axis departures easy). Either way will require some thought, since there is *no way* to make an absolute move in *X*, *Y*, and *Z* while moving incrementally in the rotary axis.

Rapid and straight-line motion. The examples given so far have shown only rapid motion commands for the rotary axis. In this mode (rapid), the rotary axis turns as quickly as it can. As stated, this mode is used for indexing only. That is, when rotated in the rapid mode, the rotary axis is being used as a simple indexer. It is simply being used to rotate the workpiece to allow machining on another side of the workpiece.

As you also know, a rotary axis can also be used to allow machining *while* the workpiece is rotating, as is commonly the case when machining certain styles of cams. A G01 straight-line motion command is used for this purpose.

You must know that something odd happens to the feed rate when a rotary-axis motion is commanded in a G01 mode. As you know, when only *X*, *Y*, and/or *Z* are involved in a G01 command, the feed rate is input in inches per minute (IPM) or millimeters per minute. Calculations for per minute feed rates are very simple to make.

Unfortunately, whenever the rotary axis is involved with a cutting command, feed-rate calculations are not so easy. Since the control has no way of keeping track of the tool's cutting tip position relative to the center of rotation, feed rate can no longer be specified in inches per minute or millimeters per minute. Whenever a rotary motion is included within *any* G01 command, the feed rate *must* be input in degrees per minute (DPM). Even if an *X, Y,* and/or *Z* motion is included within the rotary-axis command, the feed rate must still be specified in degrees per minute.

An F word is still used to designate the feed rate. The CNC control will automatically know that the feed rate is in degrees per minute if a rotary axis departure is included within a G01 command. For example, in the command

   N125 G01 X4.5 Y1.75 F5.0

the feed rate will be at 5.0 IPM, since this command includes no rotary-axis departure. However, in the command

   N125 G01 X4.5 Y1.75 B45. F5.0

the feed rate will be taken as 5.0 DPM. For all intents and purposes, the value of the feed rate in degrees per minute for a given machining operation will *never* match that for the inches-per-minute feed rate. A conversion calculation to degrees per minute must always be made.

It can be a little difficult to calculate the DPM feed rate from the feed rate required in IPM. The formula we will give is relatively easy to visualize. It involves calculating the length of cut and the machining time based on the IPM feed rate. Then, the DPM feed rate can be calculated.

Let's give a relatively simple example to help you visualize feed rate in DPM. Say you need to mill a workpiece at a feed rate of 10.0 IPM. Say the length of cut for this operation is precisely 10 in. In this case, the time required to perform this machining operation would be precisely 1 minute (a 10-inch-long cut at 10 IPM would take 1 minute). Say this operation involves only the rotary axis, and that the angular departure needed to make the cut is precisely 90°. What would be the required feed rate in DPM? Think about this before reading on.

Since the milling operation will take precisely 1 minute, and since the angular departure is precisely 90°, the required feed rate for this operation will be 90 DPM (F90.0).

To precisely calculate the desired feed rate in DPM prior to running the program requires that you apply the following formulas:

$$\text{RPM} = 3.82 \times \text{SFM} \div \text{tool diameter}$$

$$\text{IPM} = \text{inches per revolution} \times \text{RPM}$$

$$\text{Time (minutes)} = \text{length of cut} \div \text{IPM}$$

$$\text{DPM} = \text{incremental rotation amount} \div \text{time (minutes)}$$

As you can see, these formulas require that you first calculate the actual length of cut needed for machining. Unfortunately, this can sometimes be difficult to determine for rotary motions.

With the length of cut approximated, calculate the length of time required to make the cut at the desired IPM feed rate. Finally, calculate the DPM feed rate.

While we would agree that this is a somewhat cumbersome set of formulas just to calculate the proper DPM feed rate, it is important to understand how to do this, since it is required *every time* a cutting motion involving the rotary axis is required.

Because DPM feed rate is somewhat difficult to calculate, many programmers cheat. They approximate the DPM feed rate and wait until the part is being machined. Then they fine-tune the feed rate by adjusting the feed rate override switch. After overriding the feed rate to the desired level, they change the programmed feed rate accordingly.

For example, say the programmer arbitrarily selects a feed rate of 100.0 DPM for a milling operation. At the machine and during the program's verification, the programmer brings the feed rate override switch to a very low setting. As the milling operation begins, the feed rate is slowly increased until the programmer is satisfied. Say for example, the feed rate override switch is at the 80 percent position when the programmer is satisfied that machining is occurring properly. In this example, the programmed feed rate in the program would be changed to 80.0 DPM (from 100.0 DPM). Of course, by using this technique the programmer will have no idea of the precise feed rate in IPM the tool is machining with.

Special Note About G02 and G03

Keep in mind that for most CNC controls, it is impossible to form a true circular motion within a rotary-axis command. Unfortunately, there are times when it may be necessary for you to do so, especially when machining certain styles of cams. In this case, a computer-aided manufacturing (CAM) system may be required to calculate the necessary movements.

There is a relatively new feature available on some CNC machining-center controls called *polar coordinate interpolation*. It is usually an optional feature, even when a rotary axis is available. This feature does allow a true circular motion to be generated within a rotary axis command. In essence, this feature lets the programmer "flatten out" the rotary axis to view it as a linear axis. This makes programming circular commands involving the rotary axis relatively simple. We will discuss polar coordinate interpolation in detail in the next chapter. Though our presentations will be directed toward turning-center applications for a rotary axis, the same basic points will apply to the rotary axis of a machining center.

Canned cycle usage.   Rotary-axis positions can be used (just like X and Y) as hole centerline coordinates. Every X- and Y-related function of canned cycles applies directly to the rotary axis. This allows you to easily specify that a series of holes around the outside of the round part be machined. Here is an example program that machines a series of holes around the outside diameter of a round workpiece. Figure 2.8 shows the drawing to be used for this example. Note that this workpiece is to be machined on a horizontal machining center, and that program zero is the center of the workpiece in X and Z. Program zero in Y is the bottom end of the workpiece.

Program:

O00001 (Program number)
N005 G54 (Select coordinate system)
N010 G90 S500 M03 (Select absolute mode, turn spindle on CW at 500 RPM)
N015 G00 X0 Y5. B0 (Rapid to first hole location in X, Y, and B)
N020 G43 H01 Z4.1 (Instate length compensation, rapid to first Z location)

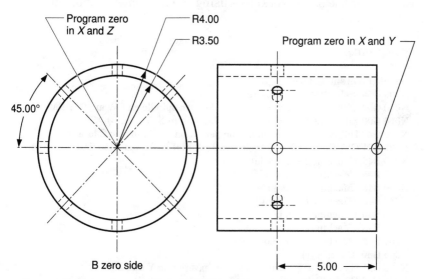

Figure 2.8 Drawing for example program stressing canned cycle usage with rotary tables.

N025 G81 R4.1 Z3.25 F4.0 (Machine first hole)
N030 B45. (Machine second hole)
N035 B90. (Machine third hole)
N040 B135. (Machine fourth hole)
N045 B180. (Machine fifth hole)
N050 B225. (Machine sixth hole)
N055 B270. (Machine seventh hole)
N060 B315. (Machine eighth hole)
N065 G80 (Cancel cycle)
N070 G91 G28 Y0 Z0 (Return to reference position in Y and Z)
N075 G28 B0 X0 (Return to reference position in B and X)
N080 M30 (End of program)

Since canned cycles are modal, a hole will be machined in each command until the cycle is canceled. Note that the motions of this program were all done in the absolute mode. For this reason, all B-axis departures were relative to the program zero point in the B axis. At the end of this program the B axis is being sent back to its reference position. In this case, it will reach the reference position by rotating another 45°. At this point, the rotary axis will be resting at a position of 360° in the absolute mode, not 0°. When this program is executed again, for the next workpiece, an unwanted full rotation will probably occur as the B axis moves counterclockwise back to the zero side!

Keep in mind that incremental techniques can be used to avoid this problem, but if they are, the canned cycle command itself would also have to be commanded in the incremental mode. Here is another program for the same workpiece using the incremental mode to specify the rotary-axis indexing.

Program:

O0001 (Program number)
N005 G54 (Select coordinate system)
N010 G90 S500 M03 (Select absolute mode, turn spindle on CW at 500 RPM)
N015 G00 X0 Y5. B0 (Rapid to first hole location in X, Y, and B)
N020 G43 H01 Z4.1 (Instate length compensation, rapid to first Z location)
N025 G91 G81 R0 Z-.600 F4.0 (Select incremental mode, machine first hole)
N030 B45. (Machine second hole)
N035 B45. (Machine third hole)
N040 B45. (Machine fourth hole)
N045 B45. (Machine fifth hole)
N050 B45. (Machine sixth hole)
N055 B45. (Machine seventh hole)
N060 B45. (Machine eighth hole)
N065 G80 (Cancel cycle)
N070 G91 G28 Y0 Z0 (Return to reference position in Y and Z)
N075 G28 B0 X0 (Return to reference position in B and X)
N080 M30 (End of program)

For most CNC controls, when canned cycles are programmed in the incremental mode, the R plane of the canned cycle command is the distance from the tool's current position in Z to the rapid plane. The Z word in the canned cycle command is the distance from the R plane to the hole bottom.

Note that since no reference is made to the program zero point in the B axis (everything is done in incremental mode), this program could be repeated over and over without any unwanted table rotation.

Keep in mind that canned cycles are but one of the special programming features that may be involved within rotary-axis commands. There may be times when knowing how other coordinate manipulation features, like mirror image, can be used within rotary-axis commands.

How to approach rotary-device programming

You have now been exposed to the raw tools used for rotary-device programming. At this point, you know the various programming commands needed to activate the rotary device. For an indexer, this simply involves specifying the proper command to make the indexer turn (usually an M code or B word). For the rotary table, you know the various commands needed to drive the rotary axis.

We will point out once again that everything discussed to this point applies equally to vertical or horizontal machining-center applications. Now we intend to present the ways you can use to approach rotary-device programs. For the most part, the principles we discuss here will apply equally to indexers as well as rotary tables and to horizontal machining centers as well as vertical machining centers.

Program zero selection.   The wise and logical selection of the program zero point in the *linear* axes (X, Y, and Z) will make programming any rotary device much easier. In fact, the key to successful and efficient rotary-axis programming revolves around the correct selection of the program zero point.

Multiple program zero points.   A logical location for program zero on one side of the workpiece may not be a logical program zero location for another side. The programmer must understand what will happen to the program zero point during an index. To begin, Figure 2.9 is a drawing that shows the front view of a vertical machining center that incorporates a rotary device.

As you can see, the rotary device is holding a fixture and the fixture is supported at one end by a tailstock. This particular fixture is being used to hold two different workpieces. It can also be used to expose more than one side of a single workpiece to the spindle for

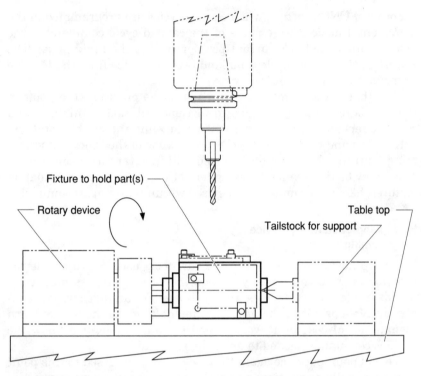

Fixture to hold part(s)

Rotary device

Table top

Tailstock for support

Figure 2.9 Front view of vertical machining center with a rotary device.

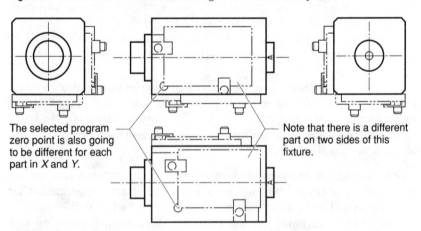

The selected program zero point is also going to be different for each part in *X* and *Y*.

Note that there is a different part on two sides of this fixture.

Figure 2.10 Assembly drawing of fixture used with a rotary device.

machining. The fixture in this example incorporates a turned diameter on the left side to accommodate the chuck jaws of the indexing device. On the right side of the fixture is a center-drilled hole to accommodate the tailstock for added support.

Now look at the assembly drawing for the fixture itself. This drawing (Fig. 2.10) is a standard orthographic projection, so you should be able to easily understand it.

This drawing more accurately describes the fact that two different parts are to be machined. The top view shows one of the parts to be machined and the front view shows the other. The end views clarify the situation yet further.

Though it may be difficult to visualize, notice that it would be next to impossible to assign one program zero point on the fixture that would make perfect sense for both parts. Figure 2.10 shows that the two parts do not share the same location points. Figure 2.11 demonstrates this more clearly, superimposing the two workpieces on the same side of the fixture.

Figure 2.11 lets you look at each side of the fixture as the spindle would see it, and makes it clear that after an index, the program zero point for the previous workpiece facing the spindle would not be the same as for the subsequent workpiece currently facing the spindle.

This is a problem the rotary-device programmer must be prepared to handle on a regular basis. One popular way to handle this problem is to assign a different program zero point for each side of the fixture. Doing this requires the use of a feature most CNC control manufacturers call *fixture offsets*. As you may know, fixture offsets allow several program zero points to be referenced from within a single pro-

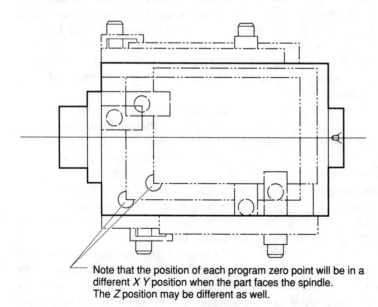

Note that the position of each program zero point will be in a different *X Y* position when the part faces the spindle. The *Z* position may be different as well.

Figure 2.11 Drawing superimposes two different sides of the fixture into one view to illustrate how program zero is affected during an index.

gram. For the previous example, two different program zero points could be easily assigned. Once assigned, the programmer could easily reference which program zero point is to be used within the program by simply specifying which fixture offset is desired.

Unfortunately, fixture offsets have several limitations. First, while fixture offsets make it much easier for the programmer to program the job, they make it more difficult for the operator to make the setup. At least three measurements (one for X, one for Y, one for Z, and possibly one for the rotary axis) must be made to determine the location of each program zero point. For this particular fixture, if one workpiece is placed on each side, at least 12 measurements would be necessary.

Second, there will be a limit to the number of fixture offsets available. For example, one popular CNC machining-center control comes with only six fixture offsets. While more can be added for an additional price, it is likely that multiple workpiece application requirements may exceed even the maximum number possible.

Third, fixture offset techniques are not available on all machining-center controls. Older controls, for example, allow only one program zero point to be designated per program.

Using the center of rotation as program zero. Another way to handle this kind of problem is to assign a central program zero point to be used by the entire program. With this method, program zero is placed in the center of rotation in two of the four axes. For vertical machining centers, if the rotary table is mounted horizontally, parallel to the X axis, program zero is placed at the center of rotation in Y and Z. For a horizontal machining center, program zero is placed at the center of rotation in X and Z. (Note that the previously shown canned cycle used this technique.)

To come up with program coordinates using this method, the programmer *must* calculate the distances from the center of rotation to each position to be commanded in the program. If fixtures are involved, this means that programmed coordinates will not match dimensions on the print. Figure 2.12 demonstrates this.

For this horizontal machining-center simple example, notice that program zero is the center of rotation in X and Z. Notice also that program zero in Y has been shifted to the lower surface of the workpiece (at least the Y coordinates will match print dimensions). For every coordinate in X and Z, the work-holding fixture must be taken into consideration. Notice that the distance from the center of index to the left edge of the workpiece in X is 2.500 in. The distance from the center of index to the back of the workpiece in Z is 3.000 in. These values were referenced for every X and Z coordinate on the coordinate sheet.

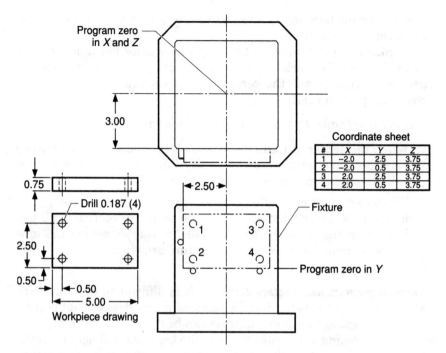

Figure 2.12 Drawing shows how program coordinates can be calculated when using the center of rotation as the program zero point.

If more than one side of this fixture holds a workpiece, the same basic techniques will be used for each side. While coordinates going into the program will not make much sense, at least the same basic techniques used to come up with program coordinates will be used for all sides of the fixture. Though this method of locating the program zero point makes it more difficult for the programmer, it provides several advantages over using fixture offsets to assign a different program zero point for each side of the fixture.

First, and most importantly, setting up the machine becomes much easier. Since there is only one program zero point to contend with, the operator has only three measurements to make. For certain applications, once the measurements have been made to the center of index one time, they need not be made again for future jobs. For example, the center of index on a horizontal machining center is the center of the table in X and Z. This position will, of course, remain the same from one setup to the next, meaning no program zero measurements will be required in X and Z from setup to setup.

Second, since program zero measurements are kept to a minimum, setup time is reduced. The amount by which it is reduced varies, depending on how many program zero points would have been assigned with fixture offsets.

Third, some machining centers, especially older machines, do not allow fixture offset techniques. *Only* one program zero point can be assigned. This means the center of index makes the only logical choice for program zero.

Other points about using the center of index as program zero

1. When the center of rotation is used as program zero, the fixture *must* be made quite accurately. Any inaccuracies in the fixture will directly affect the workpiece quality.

2. When the center of rotation is used as program zero, the programmer *must* know the distances from the center of rotation to all key location surfaces on the fixture. This, of course, means the programmer must have access to the fixture drawing.

Example program using rotary devices. It is difficult to come up with an example program showing the usage of rotary devices that is comprehensive enough to show you what can be done with the rotary axis without becoming so complicated that the beginner will not be able to follow. Here is a relatively simple example program showing a rotary table being used as an indexer. This program uses a rotary table designated as the C axis (as is the case on most vertical machining centers). However, if you have another kind of rotary device, keep in mind that there will be other commands necessary to command an index. Note that we will be using a simple G91 incremental command to make the indexes. Also note that only three sides of the rotary device will be required for the example workpieces.

This program will use four separate program zero points, assigned with fixture offsets. As you know, if the center of rotation is used as the program zero point, the coordinates being used in the program would change. As you also know, using multiple program zero points is the easier way for the programmer. But this technique makes it harder for the operator, since each program zero point must be measured.

For this example, two separate workpieces will be machined. Let's begin by looking at both prints. Note that this first workpiece (in Fig. 2.13) is marked *Large part*. The second workpiece (in Fig. 2.14) is labeled *Small part*.

Though these workpieces are quite simple, you may be surprised when you see just how long the example program is. Since rotary devices allow multiple sides to be machined, programs for workpieces held in rotary devices tend to get quite lengthy.

Figure 2.15 shows the fixture that will hold the workpieces.

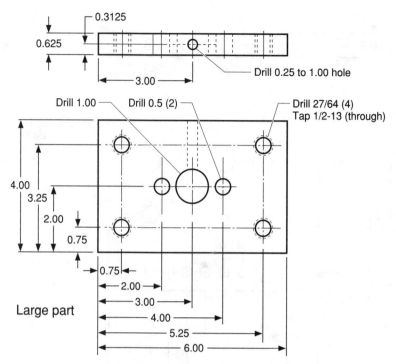

Figure 2.13   Large workpiece to be used for the example program.

Though it is made for a vertical machining center, the same techniques can be used on a horizontal machining center. Only the axes involved with the center of rotation would change.

Note from the fixture drawing that we have marked up the print to show the four locations of all program zero points. As stated, this requires that the operator make twelve measurements (X, Y, and Z × 4) during the setup.

We have also specified on the fixture drawing the fixture offsets being used for each coordinate system. The control being used for our example uses a series of G codes to specify each fixture offset. The command G54 instates fixture offset 1. G55 specifies fixture offset 2. G56 specifies fixture offset 3, and G57 specifies fixture offset 4.

For example, when our example control executes a G54, fixture offset 1 is instated. From this point in the program, subsequent X, Y, Z, and B coordinates specified in the absolute mode will be taken with reference to coordinate system 1. Whenever a G55 is executed, subsequent coordinates will be taken from coordinate system 2. And so on.

Part of preparing to write any complicated program is to prepare a sequence of operations (process). Figure 2.16 shows the sequence to

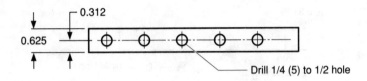

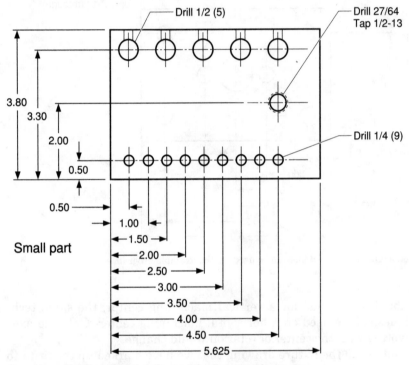

Figure 2.14 Small workpiece to be used for the example program.

be used in our example program. Notice that a tool description and cutting conditions are specified for each operation in an attempt to make it easier to follow the example program.

Note that our example machine designates the rotary axis as the *C* axis. Though the rotary axis in our setup is parallel to the *X* axis, and should be designated with the letter address A, *many* vertical machining center manufacturers designate the rotary axis in this manner.

```
O0021
N005 G91 G28 C-15. (Ensure that the plan view of large part is facing spindle)
N010 G54 G90 S1200 M03 T02 (Center drill)
N015 G00 X.75 Y.75
```

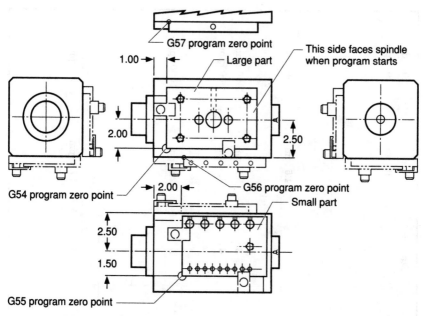

Figure 2.15 Fixture and workpieces to be used for the example program.

N020 G43 H01 Z2.
N025 M08
N030 G81 R.1 Z-.25 F3. G99
N035 X2. Y2.
N040 X3.
N045 X4. G98 (Clamp is closed, so return to initial plane)
N050 5.25 Y.75 G99
N055 Y3.25
N060 X.75 G98 (Stay above clamps for move to side view holes)
N065 G56 X.5 Y-.313 R.1 Z-.25 G99 (Note that the G56 is allowed in the canned cycle command itself)
N070 X1.5
N075 X2.5
N080 X3.5
N085 4.5 G98
N090 G80 (Cancel cycle for index)
N095 G00 Z4. (Assure clearance for index)
N095 G91 G00 C90. [Index 90° counterclockwise (CCW) to side view of large part]
N100 G90 G57 G81 X3. Y-.3125 R.1 Z-.25 G98
N105 G80
N110 G00 Z4.
N115 G91 G00 C-180. (Index 180° to plan view of small part)
N120 G90 G55 G81 X.5 Y.5 R.1 Z-.25 G99
N125 X1.
N125 X1.5
N130 X2.0

Operation	Tool	Station	Feed	Speed
Center drill holes in plan view of large part	#3 center drill	1	3.0 ipm	1200 rpm
Center drill (5) holes in side of small part				
Index 90° CCW				
Center drill (1) hole in side of large part				
Index 180°				
Center drill all holes in plan view of small part				
Drill (5) 1/2-in holes in plan view of small part	1/2-in drill	2	4.5 ipm	600 rpm
Index 90° CCW				
Drill (2) 1/2-in holes in plan view of large part				
Drill (1) 1-in hole in plan view of large part	1-in drill	3	5.5 ipm	300 rpm
Drill (4) 27/64-in holes in plan view of large part	27/64-in drill	4	4.5 ipm	600 rpm
Index 90° CW				
Drill (1) 27/64-in hole in plan view of small part				
Drill (9) 1/4-in holes in plan view of small part	1/4-in drill	5	2.0 ipm	1000 rpm
Index 90° CCW				
Drill (5) 1/4-in holes in side view of large part				
Index 90° CCW				
Drill (1) 1/4-in hole in side view of large part				
Index 90° CW				
Tap (4) 1/2-in 13 holes in plan view of large part	1/2-in 13 tap	6	17.6 ipm	230 rpm
Index 90° CW				
Tap (1) 1/2-in 13 hole in plan view of small part				
Index 90° CCW				

Figure 2.16 Sequence of operations (process) to be used for the example program.

N135 X2.5
N140 X3.0
N145 X3.5
N150 X4.0
N155 X4.5
N160 Y2.
N165 Y3.3
N170 X3.5
N175 X2.5
N180 X1.5
N185 X.5
N190 G80 M09
N195 G91 G28 Z0 M19
N198 M01
N200 T02
N205 M06
N210 G55 G90 S600 M03 T03 ($\frac{1}{2}$-in drill)
(Still on plan view of small part)
N215 G00 X.5 Y3.3
N220 G43 H01 Z.1
N225 M08
N230 G81 R.1 Z-.85 F4.5
N235 X1.5
N240 X2.5
N245 X3.5
N250 X4.0
N255 G80
N260 G00 Z4.
N265 G91 G00 C-90. (Index 90° CCW to plan view of large part)
N270 G90 G54 G81 X2.0 Y2.0 R.1 Z-.85 G99
N275 X4.0
N280 G80 M09
N285 G91 G28 Z0 M19
N290 M01
N295 T03
N300 M06
N305 G54 G90 S300 M03 T04 (1-in drill)
N310 G00 X3. Y2.
N315 G43 H03 Z.1
N320 M08
N325 G73 R.1 Z-.95 Q.1 F5.5
N330 G80 M09
N335 G91 G28 Z0 M19
N340 M01
N345 T04
N350 M06
N355 G54 G90 S600 M03 T05 (27/64 drill)
(Still on plan view of large part)
N360 G00 X.75 Y.75
N365 G43 H04 Z2.
N370 M08
N375 G73 R.1 Z-.85 F4.5 G98 (Note clamp!)

N380 X5.25 G99
N385 Y3.25
N390 X.75
N395 G80
N400 G00 Z4.
N405 G91 G00 C90. (Index 90° CW to plan view of small part)
N410 G90 G55 G73 X4.5 Y2. R.1 Z-.85 G99
N415 G80 M09
N420 G91 G28 Z0 M19
N425 M01
N430 T05
N435 M06
N440 G55 G90 S1000 M03 T06 (¼-in drill)
(Still on plan view of small part)
N445 G00 X.5 Y.5
N450 G43 H05 Z.1
N455 M08
N460 G73 R.1 Z-.8 Q.1 F2.
N465 X1.
N470 X1.5
N475 X2.
N480 X2.5
N485 X3.
N490 X3.5
N495 X4.
N500 X4.5
N505 G80
N510 G00 Z4.
N510 G91 G00 C-90. (Index 90° CCW to side view of small part)
N515 G90 G56 G81 X.5 Y-.313 R.1 Z-.5 G99
N520 X1.5
N525 X2.5
N530 X3.5
N535 X4.5
N540 G80
N545 G00 Z4.
N550 G91 G00 C-90. (Index 90° CCW to side view of large part)
N555 G90 G57 G83 X3. Y-.3125 R.1 Z-1.75 Q.5
N560 G80 M08
N565 G91 G28 Z0 M19
N570 G91 G00 C90. (Index 90° CW to plan view of large part)
N575 M01
N580 T06
N585 M06
N590 G54 G90 S230 M03 T01 (½-in 13 tap)
N595 G00 X.75 Y.75
N600 G43 H06 Z.1
N605 M08
N610 G84 R.25 Z-.85 F17.6
N615 X5.25
N620 Y3.25
N625 X.75

```
N630 G80
N635 G00 Z4.
N640 G91 G00 C90. (Index 90° CW to plan view of small part)
N645 G90 G55 G84 X4.5 Y2. R.1 Z-.85 G99
N650 G80 M09
N655 G91 G28 Z0 M19
N660 G91 G00 C-90. (Index 90° CCW back to starting point)
N665 M01
N670 G91 G28 X0 Y0
N675 T01
N680 M06
N685 M30
```

Admittedly, programs for rotary devices tend to get a little complicated. And the above program was for a relatively simple application,
since the workpieces were quite simple and only three sides of the
fixture were used. But, with some study, it will make sense.

Conclusion to rotary-device programming.   While we cannot possibly
prepare you for every possible application related to rotary-device programming, we hope your knowledge has been expanded to the point
where you can approach the rotary device programming with confi-
dence. Again, most applications require the use of the rotary device for
relatively simple indexing; it is a simple matter of rotating the workpiece to an attitude that allows machining. While these programs tend
to get quite lengthy, programming is not overly complicated.

Because it can be difficult to visualize the parts being rotated, it is
wise to make sketches and come up with a clear sequence of operations *before* the program is begun (as our example has shown). Once
the programmer can truly visualize what is going on, writing the program becomes *much* simpler.

Spindle Speed Increasers

The maximum spindle speed in RPM available from different
machining centers varies, depending on the size and application for
the machine tool itself. Many factors contribute to just how fast a
machining center's spindle is designed to rotate. If the machine is
designed to machine hard materials, and if powerful machining operations are to be performed, maximum spindle speed will probably not
be of primary concern since harder materials require relatively slow
cutting conditions. On the other hand, if the machine is designed to
machine free-machining materials with relatively fast cutting conditions, maximum spindle speed will be of much more importance.

If a company exclusively machines free-machining materials, it
will, of course, purchase a machining center designed accordingly.

However, if a company machines a variety of materials, a compromise may have to be made with regard to maximum spindle RPM. Generally speaking, a machining center that allows ultrahigh maximum spindle RPM (over say 10,000 RPM) will not have the power to perform powerful machining operations.

Spindle speed increasers allow a machining center with a relatively low maximum RPM to machine free-machining materials at (or close to) optimum cutting speeds. A spindle increaser resembles a tool holder. The cutting tool is placed in the spindle speed increaser in much the same way it would be placed in any tool holder. However, the internal workings of the spindle speed increaser are designed to increase the speed supplied to the spindle of the machine tool.

Most spindle speed increasers increase the spindle RPM by a constant factor. For example, one popular spindle speed increaser steps up the spindle speed by the ratio of 3 to 1 (3:1). For whatever spindle speed is programmed for the machining center's main spindle, this particular spindle speed increaser will multiply the speed by 3. If, for example, a speed of 3000 RPM is programmed (S3000), the cutting tool held in the spindle increaser will rotate at 9000 RPM.

As with any power transmission device, the sacrifice for allowing this increase in spindle speed is a reduction in torque. Part of this torque reduction is due to the mechanical linkages within the spindle speed increaser itself. But the major reason for torque reduction is the gearing of the spindle speed increaser. Just as the transmission of an automobile reduces torque to allow high speeds in high gears, a spindle speed increaser will never allow the same torque at the tool as is available from the spindle of the machine tool itself.

For most machining-center applications, the spindle speed indexer is designed such that it is automatically placed in the spindle by the machine's automatic tool changer. Though extremely heavy spindle speed increasers may have to be loaded into the spindle by the operator manually, most are relatively light, and can be easily handled by the machine's ATC.

Right-Angle Heads

As you know, machining centers that incorporate any kind of rotary device have the ability to rotate the workpiece and expose several sides of the workpiece to the spindle for machining. For applications when machining on more than one side of a workpiece is necessary, a rotary device is usually the best device to choose, since it allows multiple sides of the workpiece to be machined without requiring multiple setups.

However, there may be times when the cost of the rotary device cannot be justified. If, for example, a company has a limited number of workpieces that require machining on only two surfaces, the high price of the rotary device and related work-holding tooling may be more than a company is willing to pay. For these somewhat limited applications, a right-angle head provides the benefit of allowing a second surface of the workpiece to be machined without requiring the high price of a rotary device.

Since horizontal machining centers almost always incorporate a rotary device within the table of the machine, right-angle heads are seldom needed. They are more commonly required for use with a vertical machining center that does not have a rotary device. For this reason, we limit our discussions to how right-angle heads are applied on vertical machining centers. Figure 2.17 shows a right-angle head in the spindle of a vertical machining center.

As you can see, the right-angle head allows a cutting tool (an end mill in this case) to be held at a right angle (90°) to the machine's spindle. Though Fig. 2.17 shows the tool held along the X axis, keep

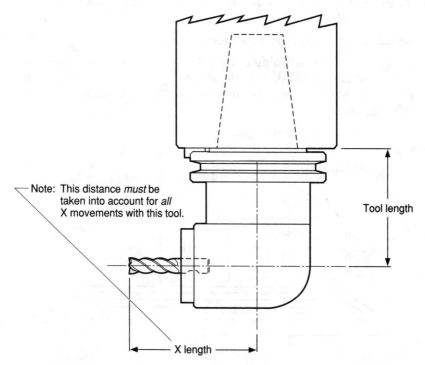

Note: This distance *must* be taken into account for *all* X movements with this tool.

Tool length

X length

Figure 2.17   A right-angle head in the spindle of a vertical machining center.

in mind that most right-angle heads can also hold the tool just as well along the *Y* axis. Also note that the tool can be pointed in either the plus or minus axis direction. This means the tool can be fashioned in a way that allows it to machine on any side around the periphery of the workpiece. Some similar tool holders even allow the angle of the cutting tool to be adjusted to a position other than 90°, allowing machining to be done on angular surfaces. For these presentations, however, we limit our discussions to right-angle heads.

From Fig. 2.17, notice that the tool length now becomes the distance from the spindle nose to the *center* of the cutting tool. The tool length compensation must reflect this distance. Notice also that the distance from the tip of the cutting tool to the center of the spindle is along the *X* axis (in this case) and *must* be taken into consideration for *all* movements the tool makes. This means the programmer must know how far the tool will be extending from the spindle centerline *before* the program can be written. This requires that the tool be preset to a precise length prior to being placed in its tool station in the machining center.

While this kind of tool can be a little difficult to program, there is a programming feature called *plane selection* which dramatically reduces the programming effort required for a right-angle head.

Understanding plane selection commands

With any vertical machining center, it is possible to machine in any one of three planes, the *X-Y* plane, the *X-Z* plane, and the *Y-Z* plane. Figure 2.18 shows the relationship of these three planes on a vertical machining center.

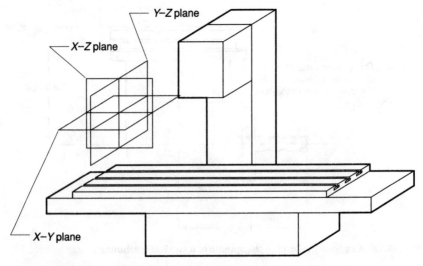

Figure 2.18   The three perpendicular planes of a vertical machining center.

Since the X-Y plane is the plane directly facing the spindle on all machining centers, almost all machining is done in the X-Y plane. For example, for machining a hole, coordinates for the hole center are usually in the X-Y plane. If milling, the path of the milling cutter is usually in the X-Y plane. Figure 2.19 shows a three-dimensional view of a workpiece as it relates to the X-Y plane.

A preparatory function (G code) is used to specify the plane in which you wish to work. On most CNC controls, G17 is used to specify the X-Y plane, G18 for the X-Z plane, and G19 for the Y-Z plane. The G17 (X-Y plane) is initialized, meaning when the power is turned on, the control automatically selects this G code, setting the X-Y plane. This means the programmer does not have to specify a G17 in the program if working exclusively in the X-Y plane.

Because so much is done in the X-Y plane, you may not have ever considered the possibility of working in another plane. And since the control will automatically select the X-Y plane at power-up, you may never have even heard of plane selection. However, when you are working with a right-angle head, knowing these implications of plane selection can dramatically reduce the programming effort. Here are some examples with extended discussions.

As stated, a right-angle head will minimize the number of secondary operations required after the CNC operation. The right-angle head can hold a variety of tools, allowing it to perform a variety of operations. Machining on the sides of a workpiece that would normally require secondary operations, can now be performed within the CNC operation. Of course, if multiple operations must be performed on the side of the workpiece, one right-angle head will be required to hold each tool.

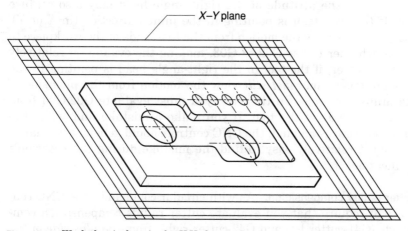

Figure 2.19 Work that is done in the X-Y plane.

Whenever a right-angle head is used, machining will *not* occur in the *X-Y* plane. It will now occur in either the *X-Z* plane or the *Y-Z* plane. Depending on the kind of operation being performed, this may make it necessary to specify the proper G code (G18 for *X-Z* plane or G19 for *Y-Z* plane) for certain CNC functions. Note that if only rapid (G00) and straight-line cutting (G01) commands will be given with the right-angle head, there is no need to specify the plane selection command. With rapid and straight-line cutting commands, the control will simply follow your series of motion commands, moving the tool to each programmed position in *X, Y,* and *Z* in a point-to-point manner.

However, there are several programming functions that require the use of plane selection when working in the *X-Z* plane and *Y-Z* plane. Let's take a look at them.

Circular commands. Any time you give a circular command, you must make the decision as to which command should be used (G02 clockwise or G03 counterclockwise) in the same way the machine will interpret the command. For *X-Y* circular moves, it's easy. Since you view the movement from the spindle nose position (the plus side of the *Z* axis), evaluating most *X-Y* circular movements is usually as simple as viewing the print from above.

However, for *X-Z* and *Y-Z* circular movements, evaluating the difference between G02 and G03 (clockwise versus counterclockwise) becomes a little more difficult. You must view the motion from the *plus side* of the uninvolved perpendicular axis. For vertical machining centers, an *X-Z* circular motion must be viewed from the back of the machine (from plus *Y*). A *Y-Z* circular motion must be viewed from the right side of the machine (from plus *X*).

Note that the attitude of the right-angle head may also confuse you. If the tool itself is pointing in the minus direction (in *X* or *Y*), you can easily view the motion from the tool's side of the workpiece to decide whether to use G02 or G03, making the evaluation relatively easy. However, if the tool in the right-angle head is pointing in the plus direction, you *must still* view the motion from the plus side of the uninvolved axis. In this case you must evaluate G02/G03 from the bottom side of the tool. If you make a mistake and select the wrong circular command, the CNC control should generate an alarm, making it relatively easy to find the mistake during the program's verification.

Cutter radius compensation. As with circular commands, the CNC control will require that you evaluate cutter radius compensation commands (G41-cutter left and G42-cutter right) from the *plus side* of the uninvolved axis. As long as the right angle head has the tool that is pointing in the minus direction (either X minus or Y minus), climb milling will be G41 (cutter left) and conventional milling will be G42 (cutter right), just as it is for machining in the X-Y plane. This assumes a right-hand milling cutter is used (spindle running clockwise). However, if the tool is pointing in the plus direction, or if left-hand tools are used, this rule for cutter radius compensation must be reversed.

Here is an example program that combines circular motions and cutter radius compensation in the same program for use with a right-angle head. Figure 2.20 is a drawing showing the workpiece to be machined. To keep this example realistic, we are machining four holes in the top surface of this part as well as milling the right side. Notice that program zero is still specified as it would be for machining on the top surface of the workpiece (at the X and Y minus corner). The end mill used to machine into the right side of this workpiece is being held in the same attitude as shown in Figure 2.17 (pointing in the X minus direction). For this example, we'll say the distance from the tip of the end mill to the spindle center line is precisely 4.000 in. As mentioned, this value must be considered for *every* movement along the X axis during the programming of the end mill.

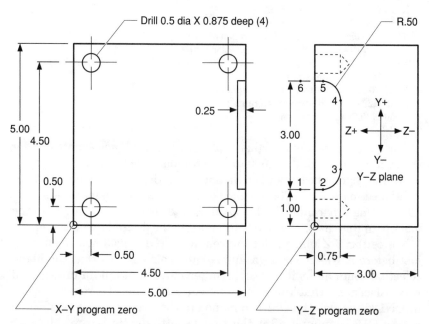

Figure 2.20 Drawing for example program using the right-angle head with cutter radius compensation.

Program:

O0016 (Program number)
(0.500-in-diameter drill)
N005 G17 G54 G90 S800 M03 T02 (Select *X-Y* plane, coordinate system, absolute mode, turn spindle on CW at 800 RPM, get tool 2 ready)
N010 G00 X.5 Y.5 (Move to first hole location)
N015 G43 H01 Z.1 (Rapid down to just above workpiece)
N020 G81 R.1 Z-.875 F5.0 (Drill lower left hole)
N025 Y4.5 (Drill upper left hole)
N030 X4.5 (Drill upper right hole)
N035 Y.5 (Drill lower right hole)
N040 G80 (Cancel cycle)
N045 G91 G28 Z0 M19 (Return to reference position, orient spindle for tool change)
N050 M01 (Optional stop)
N055 T02 M06 (Change tools to right-angle head) (0.875-in-diameter end mill)
N060 G90 S400 M03 T01 (Select absolute mode, turn spindle on CW at 400 RPM, get tool 1 ready)
N065 G00 X8.75 Y1.75 (Rapid to first *X-Y* position)
N075 G43 H02 Z.75 (Rapid down to just above workpiece)
N080 G19 (Select *Y-Z* plane)
N085 G41 D32 Y1.0 (Instate cutter radius compensation move to point 1)
N090 G01 Z-.25 (Feed to point 2)
N095 G03 Y1.5 Z-.75 R.5 (Circular move to point 3)
N100 G01 Y3.5 (Feed to point 4)
N105 G03 Y4. Z-.25 R.5 (Circular move to point 5)
N110 G01 Z.6 (Feed to point 6)
N115 G00 X9.1 (Move away in *X*)
N120 G40 (Cancel cutter radius compensation)
N125 G17 (Switch back to *X-Y* plane selection)
N130 G91 G28 Z0 M19 (Return to reference point in *Z*)
N135 M01 (Optional stop)
N140 T01 M06 (Put tool 1 back in spindle)
N145 G28 X0 Y0 (Return to reference point in *X* and *Y*)
N150 M30 (End of program)

This program requires further explanation. For all movements in *X* for the right-angle head, the 4.000 tool length (distance from tool tip to spindle center) had to be considered. In line N065, the tip of the tool had to be positioned to an *X* of 4.75. But the spindle center (of the machine's spindle) is being programmed. The spindle center is 4.000 in away from the tool tip (in this case), so it had to be added to the desired axis position.

Notice the *Y-Z* plane selection command (G19) was not given until just before it was needed (in line N080). This assures that the plane selection command will not affect other movements in an unexpected way. Also, note that the *X-Y* plane selection must be reinstated (in line N125) or else the G81 commands in the first tool will be affected by the G19 command after the first time the cycle is run. More on how plane selection affects canned cycles will follow.

Keep in mind that this tool is pointing in the X minus direction. This means that both circular movements and cutter radius compensation are evaluated from the tool point side of the motion (X plus in this case), making it relatively easy to determine which of G02/G03 and G41/G42 to use.

You may have a little trouble visualizing the motions these commands make under the influence of cutter radius compensation. You may be so used to thinking in X-Y, it may be difficult to adapt to other planes. However, the techniques shown here will be necessary if you ever have to work with a right-angle head.

Using a right-angle head with canned cycles. As you know, canned cycles allow easy programming for hole machining by machining centers. One relatively simple command is given to specify machining for the first hole, telling the control what kind of cycle to use (drill, tap, counterbore, etc.) and the necessary information needed to machine the hole (rapid plane, hole depth, feed rate, etc.). From this point, the programmer simply lists the hole coordinates at which holes are to be machined. After the last hole is machined, the canned cycle must be canceled by a G80.

There are a variety of canned-cycle types that allow the programmer to specify the kind of hole to be machined. Drilling, peck drilling, tapping, reaming, and boring are among the types most often used. Though the actual commands vary from one control manufacturer to the next, here is a list of common cycle types and their most popular G code designations:

G73 Chip-breaking peck-drilling cycle

G74 Left-hand tapping cycle

G76 Fine boring cycle

G80 Canned cycle cancel command

G81 Standard drilling cycle

G82 Counter boring cycle

G83 Chip-clearing drilling cycle

G84 Tapping cycle

G85 Reaming cycle

G86 Boring cycle

Canned cycles are most often used for machining in the Z axis. That is, as a hole is being machined, the tool is moving in the Z axis. The hole's centerline coordinates are given in the X-Y plane (G17). However, if you are using a right-angle head, hole machining will no longer be machined along the Z axis. Instead, the right-angle head will require that machining be done in the X or Y axis, depending on how the right-angle head is held in the spindle of the machine.

With the right-angle head, programming holes with canned cycles is much easier than trying to specify that holes be machined long-hand with G00 and G01 (just as it is when machining in the X-Y plane). However, as with circular commands and cutter radius compensation, you must specify the plane you intend to be machining holes in *before* you attempt to use canned cycles. Also, the meaning of each canned-cycle word within the canned cycle itself will change slightly from its meaning in the X-Y plane.

As you know, the actual words used with canned cycles will vary from one CNC control to the next. Here we show one common method of how canned-cycle usage and plane selection work together. If you understand these presentations, you should be able to easily adapt to any variation you come across. For this discussion, we will only discuss canned cycles as they are used in the absolute mode (G90).

For machining holes in the X-Y plane, the hole center coordinates are specified in the X-Y plane and machining takes place in the Z axis. Here is a list of words that are used on one popular control and their meanings when machining holes in the X-Y plane (G17):

X Hole center coordinate in X axis

Y Hole center coordinate in Y axis

R Rapid plane in Z axis

Z Hole bottom position in Z axis (work surface is program zero in Z, this word is equal to hole depth)

F Feed rate

Q Peck depth for G73 and G83

P Pause time for G82

The functions of X, Y, R, and Z will change when you change the plane selection. Say, for example, you are going to use a right-angle head as shown in Fig. 2.21. Note that the drill is pointing in the X minus direction. In this case, the Y-Z plane must be used (G19). The rapid plane (specified by R) will now be a clearance position along the X axis. The X word in the canned cycle will be the hole bottom position (along the X axis). Y and Z in the canned-cycle command will now specify the hole center coordinates.

For machining in the X-Z plane (G18), the rapid plane (specified by R) is now along the Y axis. The Y value in the canned-cycle command is the hole bottom position. The X and Z values will specify the hole centerline coordinates.

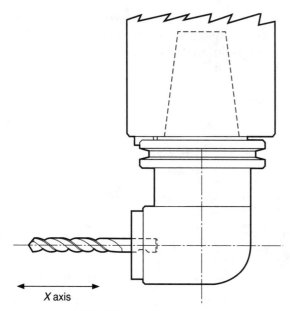

Figure 2.21 Drill held by a right-angle head.

Here is an example program showing the use of a 0.375-in-diameter drill being held in the right-angle head shown in Fig. 2.21. To keep the example program simple, say the drill tip is precisely 4.000 in from the spindle centerline. Figure 2.22 shows the workpiece to be machined. Notice that this is the same workpiece shown earlier for

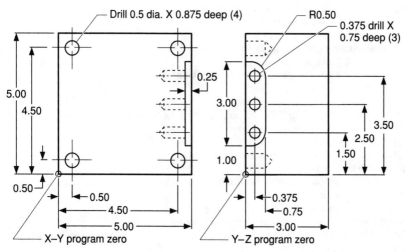

Figure 2.22 Drawing for example program using the right-angle head with canned cycles.

the circular movement and cutter radius compensation example with the addition of three 0.375-in-diameter holes. This program would, of course, require the use of two right-angle heads, one for the end mill and one for the 0.375-in-diameter drill.

Program:

O0017 (Program number)
(0.500-in-diameter drill)
N005 G17 G54 G90 S800 M03 T02 (Select X-Y plane, coordinate system, absolute mode, turn spindle on CW at 800 RPM, get tool 2 ready)
N010 G00 X.5 Y.5 (Move to first hole location)
N015 G43 H01 Z.1 (Rapid down to just above workpiece)
N020 G81 R.1 Z-.875 F5.0 (Drill lower left hole)
N025 Y4.5 (Drill upper left hole)
N030 X4.5 (Drill upper right hole)
N035 Y.5 (Drill lower right hole)
N040 G80 (Cancel cycle)
N045 G91 G28 Z0 M19 (Return to reference position, orient spindle for tool change)
N050 M01 (Optional stop)
N055 T02 M06 (Change tools to end-mill right-angle head) (0.875-in-diameter end mill)
N060 G90 S400 M03 T03 (Select absolute mode, turn spindle on CW at 400 RPM, get tool 3 ready)
N065 G00 X8.75 Y1.75 (Rapid to first X-Y position)
N075 G43 H02 Z.75 (Rapid down to just above workpiece)
N080 G19 (Select Y-Z plane)
N085 G41 D32 Y1.0 (Instate cutter radius comp, move to point 1)
N090 G01 Z-.25 (Feed to point 2)
N095 G03 Y1.5 Z-.75 R.75 (Circular move to point 3)
N100 G01 Y3.5 (Feed to point 4)
N105 G03 Y4. Z-.25 R.75 (Circular move to point 5)
N110 G01 Z.6 (Feed to point 6)
N115 G00 X9.1 (Move away in X)
N120 G40 (Cancel cutter radius compensation)
N125 G17 (X-Y plane selection)
N130 G91 G28 Z0 M19 (Return to reference point in Z)
N135 M01 (Optional stop)
N140 T03 M06 (Put tool 3 in spindle)
N145 G54 G90 S1200 M03 T01 (Select coordinate system, absolute mode, turn spindle on CW at 1200 RPM, get tool 1 ready)
N150 G00 X8.85 Y1.5 (Rapid to first X-Y position)
N155 G43 H03 Z-.375 (Rapid to hole center in Z)
N160 G19 (Select Y-Z plane)
N165 G81 Y1.5 Z-.375 R8.85 X8. F5. (Drill first hole)
N170 Y2.5 (Drill second hole)
N175 Y3.5 (Drill third hole)
N180 G80 G17 (Cancel cycle, return to X-Y plane)
N185 G91 G28 Z0 M19 (Return to reference point, orient spindle for tool change)
N190 M01 (Optional stop)
N195 T01 (Put tool 1 back in spindle)

N200 G28 X0 Y0 (Return to reference point in X and Y)
N205 M30 (End of program)

Notice once again how the plane selection commands were placed close to the actual motions requiring the Y-Z plane. We strongly recommend that you do this to avoid having unpredictable movements caused by the plane selection commands.

Notice also that plane selection commands are enclosed within each tool of the program. This means the operator could easily pick up in the middle of this program and run from the beginning of any tool.

Special Coolant Systems

Most machining centers allow at least one form of coolant system. The flood coolant system (specified by M08) is the most common machining-center coolant system and is almost always a standard feature. With a flood coolant system, one or more coolant lines are run close to the spindle. The operator adjusts these lines in a way that allows coolant to be directed at the cutting tools used in the program.

However, there are other forms of coolant system a machining-center programmer should be aware of. Though most are options, it will be helpful to know that they are available for the time when the application arises.

High-pressure coolant systems

This form of coolant system is required for certain cutting tools. When it is mandatory that the chips be washed away from the machining area, as would be the case with carbide-insert drills, a high-pressure coolant system will assure the proper chip-washing ability. Like normal flood coolant, high-pressure coolant is usually commanded by an M08.

A high-pressure coolant system tends to be a little messy, since coolant will spray in all directions around the work area. For this reason, this form of coolant system usually requires special enclosures to assure operator safety.

Through-the-tool coolant

This form of coolant system allows flood coolant to flow directly through the center of the tool, as would be needed for certain types of carbide-insert drills. Some machining centers with through-the-tool coolant allow the coolant to flow directly through the center of the machine's main spindle. This system allows relatively inexpensive tool holders to be used, since the coolant comes into the tool directly from the end of the tool. However, this kind of through-the-tool coolant system may increase machine maintenance requirements, since coolant may leak into the spindle of the machine tool.

Another form of through-the-tool coolant system pipes the coolant around the spindle. The coolant flows into the tool holder from the side of the tool. This keeps coolant out of the spindle entirely. However, the tool holder cost is higher, since a special tool holder is required for each tool requiring through-the-tool coolant within the program.

As with high-pressure coolant systems, through-the-tool coolant systems tend to be quite messy and require special enclosures to ensure operator safety.

Mist coolant systems

For certain machining operations, a better cooling effect can be accomplished by blowing on the machining operation with a combination of air and liquid. Mist coolant is also more effective than flood coolant in blowing chips away from the machining operation. When available, mist coolant is commanded by an M07.

Air-blowing coolant systems

This form of coolant is best when it is mandatory that chips be removed from the machining operation. It is seldom required on a horizontal machining center, but when, for example, pockets are milled on a vertical machining center, the chips have a tendency to accumulate around the cutting tool within the pocket. These chips will cause problems during machining. In many cases, the only way to rid the pocket of chips during machining is with a powerful blast of air.

When air-blowing coolant systems are incorporated, it is *mandatory* that there be adequate guarding to keep chips from blowing out of the machine tool itself. If they do, they could easily cause injury to anyone close to the machine.

Chip-washing coolant systems

Though this form of coolant system is not actually directed at the machining operation, it usually draws its liquid from the coolant system. For certain horizontal machining centers, chips will accumulate around the bed of the machine. If left to build up, they may eventually bind the machine's axis drive systems. Also, chips are usually quite hot as they leave the machining operation, and hot chips left to accumulate on any machine surface will play havoc with the machine's thermal stability. In order to clear unwanted chips from the machine's bed, some horizontal machining-center manufacturers use the coolant system to wash chips away.

Tap oil-shot system

In almost all cases, tapping requires that a different type of coolant be used. A thicker fluid with more lubricity is usually used for tapping. Though clearing chips from the holes to be tapped and applying tapping compound is usually the operator's responsibility, this system can be used if unmanned operation is required.

A tap oil-shot system requires a special tapping holder to be used. Like a through-the-tool coolant special tool holder, the tapping holder will be piped around the spindle. But instead of piping into the machine's coolant system, tapping compound will come from a special and separate reservoir. An M code usually activates the tap oil shot. When activated, the tapping compound will either be squirted into the hole to be tapped or onto the tap itself. Tapping compound will flow for only a short duration, just long enough to wet the tap or hole.

Keep in mind that whenever a special tapping compound is used (either applied manually or by the tap oil-shot system), it will eventually be washed away with the coolant and mix with the coolant in the coolant tank. This will contaminate the coolant and make it unusable much sooner than normal.

Rigid Tapping Systems

Most CNC machining centers require the use of a special tapping holder, called a tension/compression holder. This tapping holder is spring-loaded, and allows the tap to float in line with the spindle. The natural or neutral position for a tension/compression tapping holder is in the middle of the float travel. If the tapping holder is compressed, the tap will extend back to the neutral position when released. If it is extended, it will shrink back to the neutral position when released.

This style of tapping holder is required because most CNC machining centers do not have the ability to perfectly synchronize the spindle speed with the feed rate during tapping, especially during the spindle reversal at the hole bottom. As the spindle slows down and speeds up during reversal, the feed rate remains constant. The tension/compression holder keeps the tap from breaking, allowing the tap to rise and fall relative to the spindle itself.

Machining centers with rigid tapping (also called *synchronous tapping*) have overcome the synchronization problem. They can precisely synchronize the spindle speed with the feed rate. As the spindle slows down to reverse, so does the feed rate. As it accelerates after stopping, so does the feed rate. With rigid tapping, the tap need not be held in an (expensive) tension/compression holder.

Reduced tooling cost is but one of the benefits of rigid tapping systems. Since the machine tool can respond much faster to spindle reversals, the tapping operation itself becomes much faster. Though a machine with rigid tapping cannot quite match the productivity of a special tapping head, rigid tapping is much faster than tapping with a tension/compression holder.

When you are tapping with a tension/compression holder, the tap must be kept a safe distance away from the surface to be tapped (usually at least 0.25 in away). This is to safeguard against the possibility that the tap may pull out in its holder during tapping and be left in the hole. When you are rigid tapping, the tap can be positioned at a very close distance from the surface to be machined, since there will be no possibility that the tap will pull out in its holder. This, of course, means less travel distance during tapping and converts directly into a savings in cycle time.

Though programming techniques for rigid tapping vary from one CNC control manufacturer to the next, most have a special programming command that specifies the rigid-tapping mode (possibly an M code). Once instated, tapping can be commanded in the normal manner (usually with a G84).

Probing Devices

Probes are increasing in popularity on all forms of CNC equipment. In this chapter, we will discuss how they apply to CNC machining centers. In the next chapter, we will discuss how they apply to turning centers.

Generally speaking, probes fall into two distinctly different categories. First, they can be used as in-process gauging devices. For machining centers, the in-process gauging probe is mounted in the spindle and is used to check or monitor one or more critical features or dimensions of the workpiece. They are activated by the CNC program as part of the actual machining cycle. Probes used in this manner are generally only a small part of a more sophisticated system for automating the usage of the CNC equipment (like a flexible manufacturing system). In-process gauging probes can also be used for special purposes unrelated to measuring workpieces. Helping with tool breakage detection, finding fixture location surfaces, and digitizing a model are among these special usages.

Second, probes are utilized to assist during setup. For example, they can be mounted on the table of the machine and used to help with tool length measurements. This feature is usually called *automatic tool length measurement*. The operator can quickly and easily command that a tool's length be measured.

We must point out that there has been quite a controversy about the wisdom of incorporating any form of probing device. Indeed, there are those companies that use them constantly and swear by their use. There are other companies who will never use them and consider them wasteful.

In this section, we will begin by discussing how a common touch probe works. Then we will discuss in-process gauging probes, giving the pros and cons of using sophisticated probe devices. This will expose many of the reasons for the previously mentioned controversy and give you a way to determine whether an in-process gauging probe can help with your particular CNC situation. Next we will discuss tool touch-off probes, and show how they help during setups. Finally, we will show you many of the programming techniques used to drive a probe, giving you a good feel for what to expect as you begin utilizing any probing device.

As you will see during this last discussion, a probe can be quite a difficult device to program. For this reason, many machine-tool builders and probe manufacturers incorporate a series of previously written parametric programs in an attempt to try to make probe programming easier. During this final section related to programming, we will also show the parametric programming functions required for efficient probe use.

Admittedly, there is a great deal to talk about when it comes to probe utilization. And though we cannot hope to show you every possible aspect of how probes are used, when you are finished reading this section you should have a very good general understanding of how probes are utilized, and you should be able to determine whether a probe will help in your particular CNC environment.

How a touch probe works

Though touch probes are available in different styles, most incorporate the same basic mechanical functions. Here we will look at the basic functions. These are common to *all* probe users.

Directions of probing.   Most probes are designed to touch or probe in only a limited number of directions. The more directions of probing allowed, the more elaborate (and expensive) the probe must be. Some are quite limited in this regard, allowing only one direction of probing. A tool-length measuring probe, for example, mounted to the table of a vertical machining center (as shown in Fig. 2.23) may

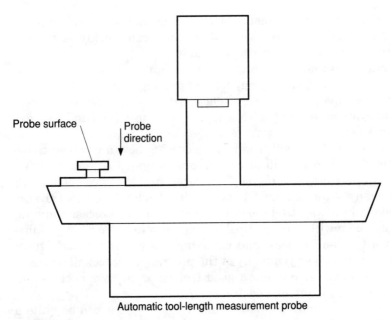

Automatic tool-length measurement probe

Figure 2.23 A tool-length measuring probe mounted to the table of a vertical machining center.

be designed to sense motion in only one direction (the Z minus direction).

The most elaborate touch probes are those used as in-process gauging devices. Most allow a total of five directions of probing. For machining-center in-process gauging touch probes (as shown in Fig. 2.24), these directions include X plus, X minus, Y plus, Y minus, and Z minus. Some even allow a combination of probing directions. However, applications that demand simultaneous multiaxis probing are few and far between.

What happens during probe contact.   Though the mechanical design of a probe will change from one manufacturer to another, most incorporate some form of spring-loaded contact device. The normal position of the probe's stylus is the neutral or undeflected condition, and in this condition the switch internal to the probe will be in an off state. As soon as the stylus is deflected to *any* extent, the switch within the probe is turned on. This switch activates a signal that is sent to the CNC control commanding that axis motion is to be halted.

How the contact signal is sent to the control.   Some touch probes incorporate a hard-wired system for feedback to the CNC control. Two

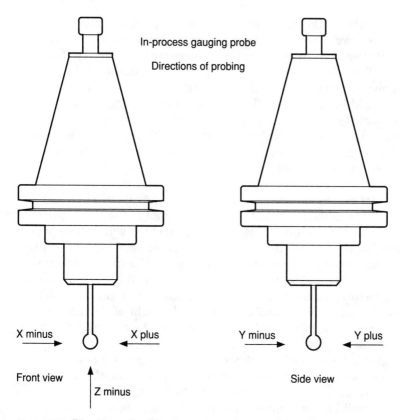

Figure 2.24 Directions of probing for an in-process gauging probe.

wires are run from the probe itself to a contact location within the CNC control. With this form of touch probe, as soon as the probe's stylus is deflected, the switch within the probe closes a circuit. The physical wires running to the CNC control complete the circuit.

With some forms of CNC equipment, like CNC machining centers, it can be somewhat cumbersome to physically run wires from the probe to the CNC control. If the machining center incorporates an automatic tool changer device through which the probe must be loaded into the spindle, the wiring required to utilize the hard-wired system becomes difficult to fabricate.

For this reason, newer probing devices incorporate a radio transmitter and receiver. The transmitter, mounted within the probe itself, is activated the instant the stylus detects a deflection. The receiver, mounted on the machine tool in close proximity to the probe, relays the signal to the CNC control.

Though the radio transmitter and receiver system has numerous advantages over the hard-wired system, there is one drawback. The transmitter within the probe must be powered by a battery. The user *must* confirm the battery quality on a regular basis. If the battery runs down to the point that the transmitter becomes inoperative, the results could be disastrous. To eliminate this possibility for disaster, the transmitter within some probes sends a constant signal to the receiver, just to confirm that the battery within the probe is still working. If for any reason the receiver does not receive this signal, it generates an alarm condition.

Allowing for overshoot. As mentioned, as soon as the probe's stylus is deflected, the probe immediately sends a signal to the CNC control to halt axis motion. Unfortunately, no CNC machine will be able to stop axis motion the instant the stop signal is detected. There will *always* be a certain amount of overshoot while the axis motion decelerates to a halted condition. The amount of overshoot will be determined by several factors, including the rate of motion (feed rate), the response time of the servo motor, and the mass and weight of the moving axis components. Generally speaking, the heavier the axis and the faster the feed rate, the more the overshoot that will occur.

The greater the overshoot, the more inconsistency there will be during probing. To minimize the amount of overshoot, most probe manufacturers recommend that the probe contact each surface twice. First the probe will be commanded to contact the surface at relatively fast rate of motion, say at about 30 to 50 IPM. This quick probing motion will simply find the approximate location of the surface. The probe is then retracted a very small amount (just larger than the overshoot amount). Usually about 0.010 in of retract motion is sufficient.

Second, the probe is commanded to contact the surface once more, at a very low rate of motion (about 1.0 to 3.0 IPM). Since the extent of this second probing action will be under 0.010 in length, it will occur quite quickly, even with the very slow feed rate. This probing technique allows very accurate, yet relatively fast, probing to be done.

Keep in mind that no matter how low the feed rate, there will *always* be a certain amount of overshoot. Also keep in mind that there will usually be a different amount of overshoot for each axis of the CNC machine tool itself. There may even be a different amount of overshoot within each axis, depending on the direction of motion (plus or minus). To illustrate this, think about the Y axis of a horizontal machining center. The components making up the Y axis of a horizontal machining center are usually quite heavy. The machine must lift the headstock of the machine whenever moving in the Y

plus direction. This motion requires a great deal of force. However, when moving in the Y minus direction, the weight of the Y-axis components tends to help with the motion. Though some horizontal machining centers have a counterbalance for the Y axis to counteract this problem, the amount of overshoot will usually be larger when probing is done in the Y minus direction.

This overshoot (no matter how small) *must* be taken into consideration *every time* a surface is touched if accurate probing is to be done. For this reason, most probe manufacturers incorporate some way of calibrating the probe to determine the amount of overshoot for each axis and axis direction. Some probe manufacturers make it easier than others to accomplish the probe's calibration. Most incorporate some form of calibration tooling as well as previously written probe calibration programs.

For the X and Y axes of a vertical machining center, for example, a ring of known diameter may be used. A probing routine is executed to measure the inside diameter of the ring. At the completion of the probe calibration routine, the control will respond, giving the user the diameter it *thinks* the ring to be. This value is commonly stored into a tool offset so the operator can easily view its value. This measured diameter will be slightly larger than the true inside diameter of the ring (due to overshoot). The operator will simply compare the diameter shown by the control to the known diameter of the ring. The amount of radial error reflects the overshoot amount. The operator stores this known overshoot amount within the control's memory by one method or another, possibly in a tool offset.

Though calibration rings are commonly used for probe calibration in X and Y, they are not truly accurate. If a ring is used for calibration purposes, the probe manufacturer is making the assumption that both axes (X and Y) will have the same amount of overshoot, and that the overshoot will be the same in both directions (plus and minus). While this may be sufficient for many vertical machining centers, since the force to move the X axis will be about the same as required to move the Y axis, it will not be very accurate for horizontal machining centers.

For a more accurate form of probe calibration, each axis, as well as each direction the axis moves in, must be calibrated independently. This means that a separate calibration routine must be executed for each direction of probing. If the probe allows five directions of probing (X plus, X minus, Y plus, Y minus, and Z minus), as does the probe shown in Fig. 2.24, five different calibration routines must be executed. The amount of overshoot for each probing direction must also be stored within the control and must be taken into account whenever probing is done.

While this overshoot discussion has been primarily directed at in-process gauging probes, the same basic points apply to tool-length measurement probes. If accuracy is required, some form of calibration tooling and calibration programs is required to determine the overshoot amount.

As mentioned, once the overshoot values have been determined, they *must* be taken into account *every* time a surface is probed. And since feed rate determines the amount of overshoot, the *same* feed rate used for calibration *must* be used whenever final probing is done.

Allowing for the stylus diameter. If measurements are to be made in the X and Y axes, the diameter of the stylus will, of course, affect measurements and must be considered whenever probing is done in X and Y. For the Z axis, since there will be only one probing direction (in the Z minus direction), and since two probed surfaces will usually be compared to each other, the need to consider the probe's stylus diameter is not so critical.

Most probe manufacturers allow a way to input the probe's stylus diameter (or radius) for use by the probing programs. In some cases, it is stored in an offset. In other cases, it is stored in a variable (more on variables during the parametric programming discussion later). Either way, this value will be available to the probing parametric program during the execution of the probing routine.

Accuracy considerations

Most probe manufacturers are claiming probe accuracies within less than 40 millionths of an inch. While this may sound like an exceptional accuracy claim, keep in mind that they can only speak for the probe itself. They are saying that before the probe's stylus reaches a deflection of 40 millionths of an inch, it will activate the switch to halt the machine's motion. Since the switch activation is only the first step of the entire probing process, it is unrealistic to expect probing accuracies within less than 40 millionths of an inch when variations in overshoot are considered.

As you can imagine, several factors determine how accurate probing will be for a given machine tool. While these factors include the accuracy of the probe itself, there are other factors that affect accuracy in a much larger way than the probe itself.

The single largest factor is how accurately the overshoot values are calibrated. Theoretically, if they are perfect, and if the machine repeats perfectly, the probe should truly register within 40 millionths. But keep in mind that this is only theoretically. In real life, this is next to impossible to achieve.

First of all, the room temperature will affect the amount of overshoot. If room temperature varies from day to day, so will the accuracy of probing. While room temperature variations may be minute, they must be considered.

More importantly, as the components of the machine tool itself warm up, the accuracy of the probe will also vary. The size of the machine tool itself will have an effect on how much this thermal deviation will affect probing. Generally speaking, the larger the machine tool, the more the thermal change will affect probe accuracy during warm-up.

For machining centers, the accuracy of the tool changer and the tool holder also enter into the picture. If the probe cannot be placed *perfectly* into the spindle (time after time), probing accuracy will suffer.

Even the normal daily wear and tear on the machine tool will affect probing accuracy. Keep in mind that if the in-process gauging probe you intend to purchase is barely capable of measuring to desired tolerances with a new machine, as the machine wears the probe will soon be rendered useless.

No single one of these factors taken by itself may present much of a problem, but when added together, they must be considered. So, how accurate will the probe you purchase for a given CNC machine tool be? This can be a difficult question to answer. Since the probe manufacturer cannot speak for the accuracy of the machine tool, they can only guarantee the accuracy of the probe itself, say 40 millionths of an inch. Since most machine-tool builders do not actually manufacture the probe, and since the probe is considered by most to be an aftermarket device, they will be reluctant to guarantee much of anything.

The best way to attain a realistic accuracy value for any probing system is to speak directly to a current user of the probing system. If you will be purchasing a probe with a machine tool, ask to speak directly to a knowledgeable person at another company using the machine tool and probe you intend to purchase.

One last note about probing accuracy that is especially related to in-process gauging probes. One basic machining practice rule of thumb is that the accuracy and repeatability of your gauging tools should be within 10 percent of the workpiece tolerance you expect to hold. This is true of *any* gauging device.

Say, for example, you determine that the realistic accuracy of the probe you buy (once all factors are taken into consideration) is $\pm$ 0.0002 in. In this case, the minimum tolerance you could realistically measure would be $\pm$ 0.002 in (10 percent of 0.002 is 0.0002). This is *not* at all a very close tolerance.

One of the primary reasons for purchasing the in-process gauging probe in the first place is to measure the *smallest* tolerances you intend to hold. If, for example, you expect to hold workpiece tolerances of 0.0005 in, the probing system's overall tolerance must be no worse than 0.00005 in to maintain the 10 percent gauge tool rule of thumb.

In-process gauging probes

Here we will discuss the benefits and drawbacks of in-process gauging probes. This should give you an insight into how in-process gauging probes are used as well as a clear understanding of the best applications for this form of probing device.

Benefit 1: reduced operator intervention.   As stated, in-process gauging probes are totally automatic and activated by the CNC program. This means that the operator of the CNC machine will not be required to perform the inspection operations accomplished by the probe.

In-process probes are most desirable and helpful when a company is attempting to remove operator intervention from the CNC machining operation. In essence, the company is utilizing the probing device instead of having the operator inspect critical tolerances of the workpiece.

Drawback 1: slow gauging.   As compared to manual measurements taken by an experienced operator or inspector, and as compared to measuring dimensions on a coordinate measuring machine, in-process gauging devices are quite slow. To accurately measure the diameter of one single hole with a tolerance of ± 0.0005 in (on a machining center) may take as long as 2 minutes, depending on the probe and probing programs. At this rate, if numerous dimensions are to be measured, it can easily take as long to gauge the workpiece as it does to machine it! If gauging is done on every workpiece, this could conceivably double production time.

Benefit 2: more good workpieces.   Depending on how the probing program is written, it is possible that, if the probe determines that a dimension is incorrect (the workpiece is out of tolerance), the workpiece may still be corrected. Since the workpiece is still mounted in the setup, in certain situations remachining may repair the incorrect dimension.

For example, say a probe is measuring the width of a slot in a workpiece machined by a machining center. Say the slot is milled by an end mill and that cutter radius compensation is used during the milling of the slot. After machining, say the probe determines that the slot is too narrow. The probing program could command that the tool offset controlling the diameter of the end mill be changed. Once the offset is changed, the end mill can be recalled to remachine the slot. All of this, of course, will be automatic and controlled by the probing program.

Drawback 2: difficult programming. The more elaborate the probing program, the more difficult it is to prepare. As you will see later, probe programming can be quite difficult. While probe manufacturers do their best to give their users a series of canned gauging programs, these standard programs cannot possibly handle every situation that comes along. If you wish to do anything outside the realm of what the standard gauging programs do, you may have to write the gauging program yourself.

Benefit 3: automatic feedback. When in-process gauging probes are used, everything is done on-line. As mentioned earlier, this allows for the possible correction of incorrect dimensions. In-process gauging probes allow other types of automatic feedback as well.

For example, there are times when a probe can be used to detect that a tool has broken. Say a machining center user is having problems with a small drill. On a regular basis, the drill is breaking during machining. After drilling, the probe can be commanded to probe into the hole. If the hole exists, the probe will extend into the hole without contacting anything. In this case, the drill has not broken. If, on the other hand, the probe contacts the workpiece surface, the drill has broken. In the case of a broken drill, the probing program can generate an alarm, halting the cycle until the operator can fix the problem.

This is but one type of the kind of automatic feedback we are talking about. Others include: recording measured dimensions to a printer, recording statistical process control data, and utilizing the desired logic with which to control workpiece tolerances (deciding how and when to make offset changes).

Drawback 3: poor environment for gauging. One of the *major* problems with in-process gauging systems is that the machining environment is not one conducive to gauging. Most machine tools are dirty and grimy. After machining, most workpieces are left with a residue of coolant and chips. There may even be burrs left on machined surfaces that must be removed before adequate gauging can be accomplished.

To combat this problem, many companies utilize some form of automatic workpiece cleaning within the machining center *before* any gauging is attempted. At the very least, an air-blowing system of some kind will blow off the workpiece prior to gauging.

Even the storage of the probe itself within the machine setup can present a problem. If the probe is stored in close proximity to chips and coolant, it can be easily soiled or damaged while a cutting tool is machining the workpiece.

Benefit and drawback 4: improved workpiece quality—the controversy. This has been a major point of contention since the inception of the in-process gauging probe. There are those in this industry who say it is smarter to measure a workpiece *before* it comes off the CNC machine, while there may still be a chance of correcting any problems. They say that removing the workpiece may make it next to impossible to make corrections to the workpiece once the part is removed. In-process gauging facilitates this on-line gauging and, hence, potential corrections to the workpiece. This is especially true with workpieces that are very difficult to handle. Extremely large castings are examples of parts that it may be wiser to measure on-line.

There are others in this industry who say you should not use the machine that makes the workpiece to measure the workpiece. They say any CNC machine will simply follow the commands of the CNC program. If a positioning imperfection has been made during the machining of the workpiece, the same imperfection may occur during the probing operation, causing a false reading on the probe's part.

Our feeling is that, as long as the overall accuracy of the in-process gauging probe is within 10 percent of the tolerances to be held, in-process gauging is a viable solution, provided that production time is not of primary concern.

Benefit 5: facilitates automation systems and unmanned operation. Indeed, this is the original (and best) reason for using in-process gauging devices. With any form of automation system, a company is attempting to totally remove operator intervention from the machining cycle. Though the extent to which any one company will go to achieve totally unmanned operation varies, any automation system will require some form of intelligence to replace what the operator will not be available to do.

Since any probing operation and the related tasks (like workpiece cleaning) add time to the machining cycle, speed cannot be of the utmost priority when in-process probing devices are incorporated in automation systems. For this reason, in-process gauging probes are not often found on stand-alone CNC machine tools. Rather, they are more commonly incorporated as part of a more elaborate automation system aimed at removing operator intervention from the CNC environment.

Probing is almost mandatory whenever unattended operation is desired, as with automation systems and flexible manufacturing cells. If an operator will be actually monitoring each cycle, as would be the case with a stand-alone machine tool, a good CNC operator can usually outperform even the best probing devices with ease.

This is not to say that you will never see an in-process gauging probe purchased with a single CNC machine tool, or that there are no good applications for in-process probes with a single machine. However, it has been our experience that the *vast* majority of companies purchasing in-process gauging probes for use with a single CNC machine seldom (if ever) use them.

Other uses for in-process gauging probes

When an in-process gauging probe is utilized, it has the ability to do more than simply measure dimensions on workpieces. In fact, some of the very best applications for in-process gauging probes (in our opinion) have nothing to do with measuring workpiece dimensions. Here are some of these somewhat unrelated applications.

Targeting to locate the program zero point. This very good probe application applies mostly to machining centers. Certain workpieces in their rough state vary dramatically from one to the next. Large castings, for example, are notorious for this kind of inconsistency. In many cases, if a hard-and-fixed CNC program is run for each workpiece in the job, the inconsistency of rough stock may cause real problems.

Say, for example, a large casting including a cored hole is to be machined. Say the hole has about 0.100 in of rough stock to be machined (on the side). If the casting varies at all, the boring operation for the hole will not remove an even amount of stock all the way around the hole. And if the casting varies more than 0.100 in from one to another, the cored hole will not even clean up.

Probably the best way to handle this situation is to design the work-holding tooling (fixture) in such a way that it targets about the cored hole. But sometimes it is difficult (if not impossible) to fabricate hard tooling to handle the targeting task, especially with larger and heavier castings. In some extreme cases, there may even be inconsistencies from one critical surface of the casting to another; one program may not even properly machine all surfaces of the casting without relocation of the workpiece within the setup.

Sometimes the easiest and best way to handle this kind of problem is to use an in-process gauging probe to find the true locations of each critical surface. For example, the in-process gauging probe could be used to find the true center of the cored hole for each workpiece to be machined. Once the in-process gauging probe finds the true center of the cored hole in X and Y, the program zero point can be automatically shifted accordingly (usually by a fixture offset).

This is but one example of when an in-process gauging probe can help with program zero targeting. Of course, the same techniques can be used to target from the rough edges of a workpiece, the edges or center of a previously machined surface, or any other critical workpiece feature.

Some companies even use the in-process gauging probe to measure other inconsistencies in setup. If, for example, it is difficult (or impossible) to perfectly align a workpiece in the setup (making it square with the machine), the probe measures the angular amount of imperfection. The coordinate system can be rotated to match the imperfection of the setup. While this may be taking the probe's feasible usage to extreme, it shows the kind of flexibility allowed by in-process gauging probes.

Detecting tool breakage. Earlier we mentioned a simple way to use the probe to determine that a drill had truly machined a hole without breaking. This is but one example of detecting tool breakage. With a little ingenuity, a probe programmer can easily use the probe to test that almost any tool has machined properly without breaking. Though this application has nothing to do with in-process gauging, it is another popular use for the in-process gauging probe.

Testing for unwanted chips. There are many operations that leave unwanted chips on the workpiece. In many cases, these chips will get in the way of subsequent operations. One example is in tapping. If the tap drill leaves chips in the hole (as would be likely on a vertical machining center), it is probable that the chips will break the tap.

For totally unmanned operation, most companies will incorporate some form of chip-clearing device. One common style is an air-blowing system. By one means or another, the air-blowing system will clear the chips from the drilled holes, ideally. However, most chip-clearing devices work with somewhat limited success.

When it is *mandatory* that chips be totally cleared, the in-process gauging probe can be used to verify the chip clearing.

Compensating for fixture position. In essence, this is the same function as locating the program zero point on workpieces with unpredictable rough stock which was discussed earlier. However, instead of working from the workpiece itself, the probe will be touching the work-holding device. When making setups, most CNC machines require that the operator measure the distance from program zero to the machine's reference position. These distances are usually manually measured and commonly stored in fixture offsets. An in-process gauging probe can automate the procedure of locating the program zero point. Keep in mind that a probing program must be written to accomplish this, meaning this is suggested only for jobs that repeat often. However, once the probe program is written, the probe can really streamline the setup procedure.

Using in-process gauging probes for digitizing. One last usage for an in-process gauging probe we wish to mention is for automatically digitizing a three-dimensional shape. After the model to be digitized is mounted on the table, the probe is programmed to touch the workpiece at previously defined intervals. In essence, a grid of probing points is described in the X and Y axes. The probe then moves down and touches the workpiece in the Z axis.

For example, say a three-dimensional shape in a workpiece 3 in square (in X and Y) is to be digitized. Say the programmer specifies that the probe is to touch the workpiece every 0.030 in in each direction. Say the probe begins at a clearance position in the Z axis and is told to start at the lower left-hand corner (X minus and Y minus) and to work its way in the plus X direction. The probe will touch the workpiece in Z, retract, move over 0.030 in (in the plus X direction), and touch again. This will be done 100 times along the X axis, tracing a length of 3 in. Then the probe will move plus in Y by 0.030 in and repeat the series of probing operations in the X minus direction. This will be repeated for the entire 3-in square. For this example (3.0 in by 3.0 in with 0.030-in intervals) the probe would contact the workpiece 1000 times.

For every time the probe contacts the workpiece, the X, Y, and Z current position (with the probe flush with the model surface) can be attained. For each point, the probe contact point is downloaded to a personal computer (through the RS-232C communications port). The personal computer stores the series of probed points in a file on its hard drive.

Once probing is completed, the series of points memorized by the personal computer can be used to generate a machining CNC program. The computer will evaluate the grid pattern of points and fill in the empty spaces with predicted points. In essence, the computer interpolates the shape to be machined and creates a workable CNC program that will machine a workpiece to within a very close proximity to the model. The more points that are probed (the smaller the move-over probing increment), the better the computer will be able to approximate the true surface of the workpiece. Once the computer has generated the complete CNC program, it is transferred back to the machining center for machining purposes.

Unfortunately, probes used for digitizing in this manner take a great deal of time. Generally speaking, each probing motion takes from 20 to 45 seconds, depending on the probing software and the accuracy expected. For our relatively small 3- by 3-in shape, digitizing could take up to 12.5 hours of probing time (at 45 seconds per contact). To optimize productivity, most companies that employ this technique let the machine do the digitizing operation through the night, when there is no one in the shop.

Tool length measurement probes

This form of probing device is designed to help with setups, assisting the operator with tool length measurements for each tool. Tool length measurements are also required when you are making new setups. There are also times during the actual running of the job that tool lengths must be measured because of tool breakage or wear. A tool length measurement probe will help in both cases.

Though some tool length measurement probes do require quite a bit of manual intervention, most allow the operator to quickly and easily specify which tools require tool length measuring. One common form of automatic tool length measuring system, for example, utilizes a standard program which is permanently stored in memory. Since this standard program is usually quite complicated to prepare, it is written by the machine-tool builder or probe manufacturer. If written properly, the standard program is quite easy to use. The operator first manually loads the tools to be measured into the machining center's tool magazine. Then, with the previously written standard program, the operator simply specifies which tool stations require tool length measuring. The program is then activated.

Once activated, the machine will automatically load the first tool to be measured into the spindle. The machine will then rapid in X and Y to a position centered above the probe. At a very fast feed rate, the tool tip will be fed in the Z minus direction until it contacts the probe. The tool will then retract slightly and touch the probe again at a very slow feed rate. After the second time the tool touches the probe, the tool length value will be automatically loaded into a tool offset corresponding to the tool just measured. This process is repeated for each tool to be measured.

For center cutting tools, like drills, taps, and reamers, the center of the spindle happens to be at the same location as the center of the tool. For this type of tool it is *very* easy for the operator to specify the tool point location in X and Y for the tool tip. However, when certain other types of tools are used, it becomes somewhat more difficult.

An end mill, for example, that has three or four flutes may require a certain amount of shifting in X and/or Y in order to assure that only the very point of one of the flutes will touch the probe. Large indexable insert face mills will require this kind of shifting. Large boring bars present the biggest problem of all, since it may be next to impossible to predict the tip location before it is loaded into the spindle. For these kinds of tools, the operator must specify, in one fashion or another, the distance and direction from the center of the spindle to the tip of the cutting tool.

Keep in mind that the tool length measuring probe must be mounted on the table before it can be used. Once mounted, most companies do not remove it unless the work-holding setup requires the table surface occupied by the probe. For this reason, automatic tool length measuring probes are most commonly found on vertical machining centers, since few work-holding setups will require the entire table surface. Automatic tool length measuring devices are not nearly as common on horizontal machining centers, because of the comparative complexity of work-holding setups and percentage of table surface they require.

Also keep in mind that automatic tool length measuring devices, though very convenient for the operator, do require that machine time be taken during the tool length measurements, since any tool length measuring device requires tool lengths to be measured on-line. This could mean as much as 2 minutes' production time per tool (depending on the automatic tool length measuring system) would be wasted while the tool lengths are being measured. In a job-shop environment, where the operator is responsible for all functions of the CNC machine, this may be acceptable. But if production time is of utmost importance, this on-line tool length measuring device will be viewed as a time waster, and another (off-line) system of measuring tool lengths should be sought.

Programming features required for probing

Any probing device can be quite difficult to program. Much of the probe programming problem is related to accurately defining what it is a programmer needs to make the probe do. While this may sound like a basic statement, there is almost no end to what an in-process gauging probe can do. In like manner, once a probing operation is finished, there is also almost no limit to what the programmer might wish to make the machine do based on the findings of the probe. This presents the probe programmer with nearly unlimited challenges.

Fortunately, most probe manufacturers have developed a series of canned routines for common probing problems. Examples of these canned routines include finding the center of a hole, finding the center of a slot, and finding the edge or corner of a workpiece. Unfortunately, these simple routines seldom allow probe users to do all that may be necessary for their probing requirements, therefore custom programs must be written. Most probe manufacturers will supply custom probing programs for their customer's requirements on a charge basis.

To learn all that is necessary to efficiently program an in-process probing device could fill an entire text by itself. Our hope in this relatively short programming section is simply to acquaint you with the most basic programming considerations. This should at least prepare you for the undertaking and give you a much better understanding of how the probe works. Since a probe requires more programming capability than standard manual programming, let's begin by looking at the additional programming features required when probing devices are used.

Stop-motion command.   One feature required for use with any probe is a stop-motion command. (Note that some CNC control manufacturers call this the *skip-cutting command*.) This command tells the machine tool to halt as soon as the probe comes into contact with any surface. Many controls use a G31 for this purpose. The G31 command behaves much the same as a G01 straight-line cutting command. Like G01, the G31 command contains an axis departure (in *X*, *Y*, and/or *Z*) along with an instated feed rate. Unlike the G01 command, the G31 command tells the control to stop the motion the instant the control receives the signal that the probe has come into contact with a surface.

For example, the command

N030 G91 G31 X-5. F5.

tells the control to move incrementally a negative 5 in along the *X* axis at a feed rate of 5 IPM. Due to the G31, as soon as the control senses that the probe has come into contact with a workpiece or fixture (or anything else), the motion will be halted. The balance of the motion will *not* be made. This contact could occur anywhere along the commanded departure, be it within 0.010 in of the beginning of the motion or up to a total of 5 in of motion. As soon as the control senses a probe contact, motion is halted.

Depending on how the control is told to respond, if the probe does *not* contact a surface within the G31 stop motion departure, it represents an erroneous condition. If this happens, the surface is not where the programmer thinks it is. Many CNC controls will generate an alarm if the probe reaches the destination point commanded within a G31 command without actually contacting anything. (Usually the control's parameter settings determine whether an alarm will be generated in this case.)

As stated, one major limitation to be overcome is that the control cannot instantly stop the motion of the machine tool. As you know, there will *always* be some lag in time and motion from the instant the probe contacts something until the motion comes to a complete halt. The larger the machine tool and the faster the feed rate, the more the probe will overshoot. Overshoot was discussed in detail earlier in this section, and as stated, must be considered every time a probing operation is done.

Parametric programming. Probing routines can be very complicated even for seemingly simple applications. In many cases, a great deal of logic (much more than is available in standard manual programming) must be built into the probing routine. Arithmetic capability, variable capability, and the ability to attain the current axis position are often needed as well. To illustrate this, say you need to develop a probing program to measure the depth of a pocket. Simple, right? It would be simple with a common depth micrometer. Probing the depth of a pocket can be much more difficult.

To begin measuring the depth of the pocket, first the top surface in Z (at the top of the pocket) must be probed. Using normal programming functions like tool length compensation and rapid motion, the probe can be easily brought to within 0.100 in of the top surface above the pocket. Using the stop-motion (G31) command discussed earlier, the probe can then be commanded to quickly touch the top surface at a fast feed rate. Then the probe can be backed off by 0.010 in and the probing can be repeated at a slower feed rate. Here is the program to this point (note that everything to this point can be done with the relatively standard CNC commands discussed to this point):

O0001 (Program number)
N005 G90 (Select absolute mode)
N010 G00 X.5 Y1. (Rapid in X and Y to a position just to the left of the pocket which is above the top surface of the pocket)
N015 G43 H01 Z.1 (Instate tool length compensation, rapid the probe just above the top surface)
N020 G91 G31 Z-.5 F35. (Incrementally, fast feed the probe into the top surface, stop at the instant of contact)
N025 G00 Z.010 (Incrementally, rapid up just a little to clear workpiece)
N030 G31 Z-.1 F3. (Probe surface again at a slow feed rate)

At the completion of line N030, the probe will be resting flush with the top surface of the workpiece. Now what? How can you get the control to memorize the current position in the Z axis so it can be compared later to the lower surface that will be touched? Worse yet, once the probing routine *has* measured the pocket depth, how do you get the control to calculate the depth of the pocket. What will you have the control do if the pocket is too shallow? Too deep?

This simple example should begin to open your eyes, exposing some of the difficulties for programming an in-process gauging probe. Standard CNC programming commands are just not sufficient for probe programming. Probes require the CNC control to be much more intelligent than when it is activating standard CNC programs. Parametric programming gives the control this intelligence.

Just as importantly, the probe programmer *must* have a very clear understanding of what it is the probe must do. A great deal of programming time can be wasted if the probe's intended function for a given application is not completely understood. While parametric programming is not always a standard feature on some CNC controls, it *must* be used whenever a CNC machine is to be equipped with a probing device. Parametric programming must be part of the standard package whenever a probing device is purchased.

To begin to describe parametric programming, it can be easily compared to the computer programming languages BASIC or C language used with personal computers. Almost every computer-related feature of BASIC or C is available from within a parametric program. Along with these computer-related features, most versions of parametric programming also include a series of needed CNC control features that blend parametric programming into an extremely powerful probe programming tool.

Unfortunately, we could fill a text with information about how parametric programming is done. For this reason, in this section, we will discuss only those features of parametric programming that relate to probing, and only to the extent that they ensure your understanding of probe programming.

Variable capabilities. As in any computer programming language, variables within parametric programs are temporary storage locations, in which values needed by the parametric program can be stored. They are used for numerous purposes, and can be compared to the memories on an electronic calculator. With a calculator, you can store a value that is commonly needed during calculations into one of the calculator's memories. Any time the value is needed, the user can simply press one or two calculator buttons to recall the value. In much the same way, values stored in variables of parametric programming can be easily recalled for use at any time within the parametric program.

One popular version of parametric programming uses a pound sign (#) to represent variables. For example, #101 represents variable 101, #102 is variable 102, and so on. The command

#101 = 5.0

places the value 5.0 in variable #101. At any subsequent time during the execution of the parametric program, if variable #101 is evoked, it will be accepted as a value of 5.0. Variables can be used (among other things) as constants, as part of arithmetic calculations, and even to store current axis positions.

Access to the machine's current position. When *any* kind of probing is done, the parametric probing program *must* be able to determine the current position of an axis at the point of contact. This is one of the most basic requirements of probing in the first place, and is stressed in the previous pocket depth measuring example. Most versions of parametric programming use a special series of variables to represent the current position in each axis at any time. One popular version of parametric programming calls variables used for this purpose *system variables*. For example, this particular version of parametric programming uses system variable #5061 to represent the current absolute position of the X axis, #5052 to represent the current absolute position of the Y axis, and #5063 to represent the current absolute position of the Z axis. Of course, the "current" position of any axis constantly changes whenever motion is made, meaning these system variables will change with axis motion. They can easily be accessed by other variables. For example, in the previous example related to measuring the depth of a pocket, let's add one command:

O0001 (Program number)
N005 G90 (Select absolute mode)
N010 G00 X.5 Y1. (Rapid in X and Y to a position just to the left of the pocket which is above the top surface of the pocket)
N015 G43 H01 Z.1 (Instate tool length compensation, rapid the probe just above the top surface)
N020 G91 G31 Z-.5 F35. (Incrementally, fast feed the probe into the top surface, stop at the instant of contact)
N025 G00 Z.010 (Incrementally, rapid up just a little to clear workpiece)
N030 G31 Z-.1 F3. (Probe surface again at a slow feed rate)
N035 #101 = #5063 (Store current absolute Z position in variable #101)

In line N035, the current Z position (with the probe flush with the top surface of the workpiece) is being stored in variable #101. After probing the lower surface of the pocket, the (new) current value of #5063 can be compared to #101. While the probe is resting at the lower surface of the pocket, the command

N075 #102 = ABS[#5063 - #101]

will store the depth of the pocket in variable #102. (ABS represents absolute value, meaning the result of the calculation will always be a positive number.) Though this example did not take into account the overshoot of the probe, you should be starting to see how the current position can be attained and used within the parametric probing program.

Logic statements and branching. *All* probing routines require that the parametric probing program be able to make decisions. The pocket depth probing example illustrated this. After the pocket depth is measured, there is a variety of things the programmer may wish the machine to do, based on the findings of the probe. If the pocket is too deep, the programmer may wish to halt the cycle. If the pocket is too shallow, the programmer may wish to have the pocket remachined. If the pocket is to size, the programmer may wish the program to continue in the normal manner.

An IF statement (also called a *conditional branching* statement) can be used within the parametric probing program to make tests. Two variables (or arithmetic expressions) can be compared with a logical expression. If the logical expression is true, the machine can be told to do one thing. If the logical expression is false, the machine can be told to do another. This feature of parametric programming gives the probing program an ability to act on one of two possibilities, depending on whether the condition tested by the IF statement is true or false.

Here is an example of an IF statement that could be used in the previous pocket depth measuring example:

IF [#102 GT 0.500] GOTO 25

In this case, the measured pocket depth (stored in common variable #102) is being compared to its intended depth. If the measured depth is greater than (GT) 0.500 in, the condition of the IF statement will be evaluated by the control as TRUE. In the case of a TRUE condition, the GOTO statement is executed. In this case the control would jump to line N25 and continue executing from there. Since this condition would represent an oversize condition of the pocket, possibly the programmer would cause an alarm to be generated in line N25.

If the pocket is equal to or less than 0.500 in deep, this IF statement would be evaluated as a FALSE statement. If an IF statement is evaluated as FALSE, the control will simply continue executing the *next* line in the parametric program.

As you can imagine, the IF statement opens the door to a world of testing possibilities. Virtually any condition of probing can be tested. The machine can be easily told to behave differently, depending on the results of the test.

Access to tool offsets. Most parametric programs must have access to the CNC control's tool offsets. By incorporating tool length compensation and cutter radius compensation within the machining program, a probe programmer can easily manipulate the size of the workpiece being machined by changing the corresponding tool length or cutter radius offset.

There are two times when the ability to manipulate tool offsets is mandatory. First, many times remachining is necessary after probing if a dimension is not yet machined to size. In our simple pocket depth measuring example, say the parametric program determines that the pocket is too shallow. In this case, the probe programmer may wish to have the parametric program remachine the pocket to a greater depth. One way to accomplish this is to modify the tool length offset for the milling cutter that machined the pocket. Once the offset is changed, the program would recall the milling cutter and remachine the pocket.

Another similar time a tool offset must be changed by a parametric program is when the measured dimension is not in the middle of its tolerance band (at the mean value of the dimension). Maybe the parametric probing program determines that the pocket is not too deep or too shallow, but it is not at its perfect depth either. In this case, the parametric programmer may wish to change the milling cutter's tool length or radius offset to ensure that the *next* workpiece is perfectly machined.

This kind of offset changing can get a little sticky. Most in-process gauging probe programmers will not change an offset for every workpiece. If this were attempted, the dimensions of the machined workpieces will tend to bounce from one end of the tolerance band to the other. Instead, most follow the rules of statistical process control (SPC), and allow dimensions of the workpiece to follow a predetermined dimension trend.

One popular version of parametric programming uses a series of system variables to represent tool offsets. System variables in the 2000 series (#2001–#2099) are used for this purpose. System variable #2001, for example, contains the current value of offset 1. System variable #2002 contains the current value of offset 2. And so on. These tool offset system variables can be read as well as overwritten.

In the pocket depth measuring example, say the milling cutter used for machining the pocket uses offset 5 for its length compensation offset. After probing, say the pocket is measured to be shallow. Say that the parametric programmer makes variable #101 equal to how much too shallow the pocket currently is. As long as the tool's length is stored in offset 5, and as long as #101 is a positive value, the following command would modify offset 5's value by the correct amount:

N105 #2005 = #2005 - #101

After this command is executed, the milling cutter can be recalled by the parametric program and machining will reoccur. Based on the new value of offset 5, the pocket will come out on size.

Alarm generation. There are many times during probing when the parametric programmer must plan for problems. When problems occur, the programmer can actually generate an alarm to help diagnose the problem. For example, in the pocket depth measuring example, say the pocket is measured to be too deep. In this case, it is probable that the workpiece is already scrap and to continue with the program would be pointless. Or possibly, the top surface could be remachined to make the pocket shallower. In either case, machining must be halted.

One popular version of parametric programming uses system variable #3000 to generate alarms. If read, the #3000 command will put the machine in an alarm state and display a message. Here is an example:

#3000 = 150 (POCKET IS TOO DEEP!)

If this command is read, the control will halt the cycle and display the alarm message "POCKET IS TOO DEEP!" on the display screen. The alarm number 150 is determined by the parametric programmer and allows further documentation related to the alarm to be developed separately. Possibly the parametric programmer will develop a special alarm list for the alarms his or her programs generate.

Example probing parametric program. With the previous points understood, you should be able to easily follow this parametric program to measure the depth of a pocket. While you may not be yet ready to tackle more difficult probe-programming tasks, we hope this information has broadened your knowledge of probing.

To keep the example as simple as possible, let's set some basic and simple criteria. (By the way, setting the basic criteria for what you intend the probe to do is *always* the first step to preparing a probing program.) We'll say the probe has already been positioned in Z to a location 0.100 in above the top surface and just to the left of the pocket in X. The parametric program will continue from this point. We'll also say the pocket is supposed to be 0.500 in deep, + 0, − 0.005 in. If the pocket is not to size, the example program will generate a specific alarm. If it is anywhere within the specified tolerance, the program will simply continue.

We'll also say that the overshoot calibration value for the Z axis is stored as a plus number in permanent common variable #503. This is the value that must be taken into consideration whenever probing is done along the Z minus direction.

Note that most parametric programs do not use sequence numbers for every line. We do include them only for documentation purposes. This parametric program is in one popular format called Custom Macro version B.

```
O1000 (Program number)
N005 G91 G31 Z-.5 F35. (Fast feed to contact top surface)
N010 G00 Z0.010 (Feed back a small amount)
N015 G31 Z-.1 F3. (Slowly touch top surface)
N020 #101 = #5063 - #503 (Store current position in #101, allow for overshoot)
N025 G00 Z.1 (Rapid up to clear surface)
N030 G00 X.75 (Incrementally, move above pocket)
N035 G00 Z-.5 (Rapid down to just above pocket bottom)
N040 G31 Z-.5 F35. (Touch pocket bottom, note still incremental)
N045 G00 Z0.010 (Feed back a small amount)
N050 G31 Z-.1 F3. (Slowly touch pocket bottom)
N055 #102 = #5063 - #503 (Store current position in #102, allow for overshoot)
N060 G00 Z.6 (Rapid up above top surface)
N065 G91 G28 Z0 (Return to reference position in Z)
N070 #103 = ABS[#102 - #101] (Calculate pocket depth)
N075 IF[#103 LE 0.500] GOTO 085 (Test pocket depth)
N080 #3000 = 150 (POCKET TOO DEEP) (Pocket is too deep, generate alarm)
N085 IF[#103 GE 0.495] GOTO 095
N090 #3000 = 151 (POCKET TOO SHALLOW) (Pocket is too shallow, generate
alarm)
N095...(Program continues)
    .
    .
    .
    .
    .
N200 M99 (End of parametric program)
```

With study, this parametric probing program should make sense. Admittedly, it is rather limited. Our intention was simply to show you the most basic functions of probing.

Conclusion to probing devices

As you have seen, there are numerous considerations when it comes to probing of any kind. We hope we have shed some light on this somewhat confusing subject. At the very least, you should now have a basic understanding of the probe types, their applications, and the related programming techniques needed to take advantage of these sophisticated devices as they apply to CNC machining centers.

Tool-Life Management Systems

Production quantities vary dramatically from one company to the next and even from one job to the next. For low production quantities, it is likely that one set of cutting tools will completely machine the entire set of workpieces. For example, if you are machining only 10 workpieces, and if cutting conditions have been programmed correctly, it is unlikely that any of the tools will dull before the workpieces are completed.

As production quantities grow, however, so does the likelihood that one or more of the tools will dull before the entire job is run. In most cases, the operator simply stops machining workpieces just long enough to load a new tool and enter its new offset values. For relatively low- to medium-sized production quantities, this may be an adequate solution to the tool wear problem.

Keep in mind that with the previous example, whenever the operator is loading new tools and entering new offset values, the machine cannot be producing workpieces. With high production quantities, the downtime suffered due to tool replacement can become a substantial factor that slows production. The higher the production quantity, the more substantial the tool replacement problem will become, and the more the need for a company to do something to solve it.

Most CNC machining-center tool magazines are designed to hold far more tools than are needed by the typical CNC program. Say, for example, a particular CNC machining center holds 30 tools. Say that one program used to machine a workpiece with high production quantities requires only 10 tools. In this case, there will be 20 tool stations not involved with the CNC program. These tool stations can be easily used as duplicate tool stations to double or triple those tools that are most prone to dulling during the machining cycle.

In almost all programs, there will be those tools that are more prone to dulling than others. For example, a large rough milling cutter used to machine a great deal of material from the workpiece will dull within a smaller number of workpieces than a 0.500-in drill used to machine only one hole. For those tools that are most prone to wear, two or more identical tools can be placed in the machine's tool magazine to minimize the number of times the tool must be changed during the machining cycle. This technique is known as *tool-life management.*

A simple tool-life management system

How tool-life management is done depends on how much effort and expense the company is willing to invest in reducing the tool replacement problem. This first method, though cumbersome, is used by companies not willing to purchase special options related to tool-life management. It incorporates simple and standard CNC programming techniques.

Say, for example, the programmer determines that a rough milling cutter is dulling at a much faster rate than all other tools in the program. After only 50 workpieces, for example, this milling cutter becomes dull. The rest of the tools in this job last for 150 workpieces. For this reason, the programmer decides to use three identical cutters for the job, one placed in station 1, one in station 21, and one in station 31.

For programming purposes, three separate, yet extremely similar, CNC programs will be placed in the control's memory. The only differences in these three programs will be the tool station number and offset numbers for the rough milling cutter. These programs will be stored as subprograms and be automatically executed 50 times each. Say program O1001 uses the rough milling cutter in station 1, O1002 uses it in station number 21, and O1003 uses it in station number 31. Here is the basic structure of this programming technique for one popular CNC control.

```
O1000 (Main program number)
N005 M98 P1001 L50 (Machine first 50 workpieces)
N010 M98 P1002 L50 (Machine second 50 workpieces)
N015 M98 P1003 L50 (Machine third 50 workpieces)
N020 M30 (End of main program)
```

This kind of tool-life management system is commonly used to simulate what a true tool-life management system is designed to do. Unfortunately, it has several limitations. First, as the number of tools that must be managed grows, this technique becomes much more difficult to work with. Though our example may have seemed quite simple, keep in mind that only one tool needed to be managed (the face mill). Also, we set an arbitrary number of workpieces that all the rest of the tools could machine without dulling (150). Since each tool in the program will remain sharp for a different length of time (and number of workpieces), managing more tools in the real world becomes quite a bit more complicated.

Second, this system may not truly reduce the time required for tool replacement—it may just postpone it. In our previous example, eventually all three end mills will have to be replaced. If over 150 workpieces are required, the machine will sit idle while new tools are reloaded.

Third, this technique makes it next to impossible to restart in the middle of the entire cycle. If, for example, the operator must stop the cycle for any reason (lunch break or end of shift), the cycle cannot be continued where the operator left off.

For these reasons, most companies that wish to take their tool-life management system seriously need more than basic CNC programming techniques. They need functions designed specifically by the CNC control manufacturer to help with tool-life management. Unfortunately, many times the control manufacturer will make the tool-life management feature an option that must be purchased at an extra cost.

Control-manufacturer-designed tool-life management systems

These systems vary dramatically from one CNC control manufacturer to the next. While we intend to be quite specific, and the specific techniques we show are for one popular CNC machining-center control, you must be prepared for variations.

As mentioned earlier, tool-life management systems are usually used for high production, when it is likely that a tool will dull long before all production is run. Keep in mind that these elaborate tool-life management systems are also required as part of more sophisticated flexible manufacturing systems and automation systems. With these systems, the basic goal is unmanned operation. Tool-life management systems are often required if this goal is to be achieved.

Understanding tool groups.   Most programmers are used to thinking of single tools as performing machining operations. As you know, the T word is commonly used to specify a tool station number. In each tool station resides a single tool. T01, for example, commands tool station 1. Only one tool can be in tool station 1. In similar fashion, the H word (on many CNC controls) is used to designate a single tool offset, the one to be used to specify the length of the tool. In like manner, the D word designates the offset used to specify the tool's radius during cutter radius compensation commands.

With any form of tool-life management supplied by a CNC control manufacture, the functions of these elementary programming words will change. A programmer can no longer view a machining operation as being performed by a single tool, but instead must view the operation as being performed by a *tool group*. Of course, all tools in any one group will be of identical configuration and each must be capable of machining with the same cutting conditions.

Though CNC controls vary with regard to how many tools they allow to be placed in any one group, most allow at least 10. One CNC control we know of allows up to 99 tools to be utilized in one tool group.

Within each tool group, the CNC control will allow the operator to easily specify which tool stations, H offsets, and D offsets are related to each tool in the group. For example, say tool group 1 is for a rough milling cutter. Say the operator intends to place identical rough milling cutters in stations 1, 2, and 3. In this case, group 1 will include three tools, and the operator will tell the control that tools 1, 2, and 3 are to be the tools in the group. The operator will also tell the control that offset 1 is the length offset (H) for tool 1, and that offset 31 is the radius offset (D) for tool 1. In like manner, the offsets related to tools 2 and 3 must be specified.

All of this is done within group 1. If more tool groups are required, the operator will enter the data corresponding to each group accordingly. Note that, since this tool group data entering does take time, most CNC controls even allow the tool group data to be entered by programmed commands, eliminating the need for the operator to do so manually.

Programming for tool groups.   As stated, when a tool life management system is used, the programming of the T, D, and H words will change slightly. The T word is no longer simply the tool station number. To avoid confusion with normal tool station numbers, one popular control uses a series of T words in the 100 series to command tool groups. For example, T101 specifies tool *group* 1, T102 specifies tool group 2, and so on. Keep in mind that there will be absolutely no relationship between the tool group number and the station numbers in the group. Tool group 1, for example, may include tool stations numbered 19, 22, 23, and 25.

The D and H words used to specify tool offsets related to each tool in the group must also change. Since the tool station number related to each tool in the group may be unknown to the programmer at the time of programming (as may be the quantity of tools in the group), he or she may not know what offsets are to be related to the tools in the group. For this reason, one popular CNC control uses D99 and H99 to specify that the offsets given within each tool group and for each tool must be used. Here is a portion of a CNC program that stresses this.

O0001 (Program number)
N005 T101 M06 (Place the active tool in group 1 into the spindle)
N010 G90 S500 M03 (Select absolute mode, start spindle CW at 500 RPM)
N015 G00 X-.6 Y-.6 (Rapid to first X and Y position)
N020 G43 H99 Z-.6 (Instate tool length compensation for the active tool in the tool group)
N026 G01 G41 D99 X0 F5. (Instate cutter radius compensation for the active tool in the group)
...

As you can see, no hard-and-fixed tool stations or offsets are referenced during programming. Only the tool group number is given.

D99 and H99 specify that the offset related to the active tool be used. The control will automatically figure out which tool is active and adjust accordingly.

Controlling the life of each tool within the group. Most CNC controls allow the user to control the tool life in one of two ways. One way is to specify the total number of workpieces each tool can machine. By this method, information within each tool group will change from setup to setup, since the total machining time per tool will vary from one job to the next.

A better way to specify tool life is in machining time. With this method, tool life data may not have to be changed for popular tool groups from one job to the next, as long as similar cutting conditions are used.

Keep in mind that, though not often required, the operator will usually have the ability to specify a different tool life value for each tool within a tool group. This will be helpful if slightly different styles of tools are used within the group. For example, the company may own both standard high-speed steel (HSS) as well as cobalt end mills. The cobalt end mills may last longer than the HSS end mills, in which case they can be assigned a longer life. Keep in mind, however, that all tools in the group must run at the same speeds and feeds for most versions of tool-life management.

Monitoring and updating the tool-life management system. The operator will easily be able to monitor the status of each tool group. From the tool-life management display screen, the operator will be able to tell which tool stations are related to each group, which tool station is currently the active station within each group, and at what point in its life (current cycle time or workpiece count) the active tool happens to be.

As production is run, tools within each tool group will eventually become dull. The tool-life manager (within the CNC control) will automatically move on to the next tool in the group. At this point, the CNC control will tag the exhausted tool on the tool-life manager display in a way that will make it obvious to the operator that the tool is no longer usable. The next tool in the group will become the active tool. It will be the tool selected the next time the tool group is specified during a tool-change command.

At some point, the operator will replace the exhausted tools in the tool magazine. Most machines with tool-life management allow this to be done while the machine is running production so that no production time will be lost during tool replacement. Once a tool has been replaced, the operator must remember to enter the new tool's offset information and update the tool-life manager. This is easily accomplished through the offset display screen and tool-life management display screen. Keep in mind that if all tools within a group become exhausted, an alarm will sound and the cycle will be stopped the next time the tool group is commanded during a tool change.

Tool Breakage Detection Systems

All cutting tools used on a machining center will eventually dull. If left to continue machining unchanged, they will eventually break. Theoretically, if cutting conditions are properly commanded for any cutting tool, the amount of machining time they will allow for a given workpiece material should be quite consistent. If so, it will be relatively easy to determine how long each tool will last. If a programmer can predict how long a given tool will last (or if this can be determined through testing), a tool-life management system can easily ensure that tools used in a program will never dull to the point at which they will break. In this case, no tool breakage detection system would be necessary.

Unfortunately, there may be certain cutting conditions that vary from one workpiece to the next. For example, the hardness of the workpiece material must remain quite consistent in order for tool life to remain the same from tool to tool. If the hardness of the workpiece material varies, optimum cutting conditions for a workpiece of one hardness will not be correct for another. Castings and forgings are notorious for their hardness inconsistencies from one workpiece to the next.

Another workpiece material variation that will cause havoc with machining operations comes from imperfections within the workpiece itself. Again, castings are known for these kinds of imperfections. The outside surface of a casting has a certain amount of flash or crust that is harder than the material within. While flash can be removed by milling cutters to allow surface machining, keep in mind that flash will occur on all surfaces of the casting. During the casting process, if an air bubble forms inside the casting, the surface around the bubble (inside the casting) will be much harder than the bulk of the casting itself. If a cutting tool enters the area of an air bubble, it will immediately be under a great deal of stress. At best, the tool will dull quickly. At worst, it will immediately break.

Another casting-related problem is that during the casting process, it is possible that foreign materials will be cast with the casting material. Since castings are often made from the scrap of previously machined workpieces, often the raw materials for the casting are contaminated. For example, it is not uncommon for carbide inserts to be thrown into the chip barrel with the chips from a machining operation. These chips, of course, will be used again to make future workpieces. If there are foreign materials within the casting, and if a cutting tool comes into contact with them, it is likely that the tool will immediately dull and break.

The last material variable we will mention is the amount of rough material that must be removed. If the amount of stock varies from one workpiece to the next, tool life will also vary.

As you can see, there will be many times when even a good tool-life management system will have limitations with regard to consistently predicting the true life of a tool. Moreover, if a tool dulls and breaks during the machining cycle, and if the operator does not notice it (as would be the case in an unmanned system), the results could be disastrous.

In many cases, tools within a program depend on one another. A tap, for example, requires that a tap drill be used prior to the tapping operation. If the tap drill breaks, and if left unchecked, of course the tap will eventually break as well. Other examples of this dependency from one tool to another include rough and finish milling cutters as well as rough and finish boring bars.

A tool breakage detection system is designed to monitor tools to assure that they are still capable of machining. These systems vary dramatically with regard to what kinds of tools are to be monitored as well as what will happen if tool breakage is detected. By no means do we wish to imply that the forms of tool breakage detection systems we show here are the only forms available, since the technology of these systems is constantly changing.

Tool breakage detection systems fall into two basic categories. One monitors horsepower draw and the other probes each tool after machining. The best tool breakage detection systems incorporate both of these methods.

Horsepower-monitoring tool breakage detection systems

One system utilizes the spindle power draw with which to detect a tool's degree of distress. With this system, the tool breakage detection system is "taught" how the tool will normally machine the workpiece. During this teaching process, a workpiece is machined in the normal manner. The tool breakage detection system monitors and records the cycle, finding the peaks and valleys during the program cycle.

After the tool breakage detection system has learned how the cycle should run, the operator specifies the degree of inconsistency (usually in percentage) that the system can accept before a tool breakage is determined. For example, the operator may specify a degree of inconsistency of 10 percent. If, during the machining cycle, the system finds a horsepower draw increase of more than 10 percent (for that point in the program), it decides that a tool has broken.

This system is most often used for large, powerful machining operations like face milling. With these kinds of operations, larger tools are usually involved which allow a greater percentage of the machine tool's horsepower to be used.

It is infeasible to use this kind of system for small tools which use only a small percentage of the machine's horsepower. When a small tool is machining, say a $\frac{1}{8}$-in-diameter drill, the system may not even register the fact that a machining operation is occurring.

Variations of this kind of tool breakage detection system monitor the draw on axis drive motors as well as the spindle drive motor. Since one or more axis drive systems will be driving the tool into the workpiece, they can also be monitored to determine a tool's degree of distress.

The main benefit of this style of tool breakage detection system is that it is dynamic, monitoring for tool breakage as the machine is actually machining a workpiece. If tool breakage is detected, the machine will stop before serious damage can occur to the tool and workpiece.

Unfortunately, this style of system also has its limitations. Since only motor horsepower (that of spindle and possibly axis drive motors) is being monitored, the system will have no way to distinguish if a rise in horsepower is being caused by tool breakage or a change in material consistency. If, for example, there are material variations in hardness or stock amount, the percentage of inconsistent power draw may have to be set to a large value, rendering the system useless except for catastrophic problems.

Probing-style tool breakage detection systems

This form of tool breakage detection system utilizes a probe to check whether a tool has broken during its machining operation. The probe can either be mounted in close proximity to the tool that is in the spindle or be mounted close to the tool changer magazine. Either way, the probe is used to check whether the tool's point is still intact.

During the tool checking, the tool tip is brought into contact with the probe. If the tool point actually touches the probe, the system determines that the tool is still capable of machining. If the probe does not sense the tool point, the tool is determined to be broken.

This form of tool breakage detection system is best for small tools like small-diameter drills and taps. For example, the tap drill can be tested for breakage before the tap is allowed to machine. If the drill has broken along the way, the cycle can be stopped before the tap is run.

Keep in mind that if the tool in the spindle is being checked, tool breakage testing will require machining cycle time to be taken during the test. For this reason, some probing-style tool breakage detection systems are mounted along the automatic tool-changer magazine. While this form of tool breakage detection is usually more expensive, since the probe must have some way of adjusting for tool length, at least tool breakage detection tests can be made off line, and not during the machining cycle.

When tool breakage is detected

Most tool breakage detection systems allow the user to specify what is to happen if a tool breaks during the machining cycle. One common choice is to stop the machine until the operator can remedy the situation. In this case the machine will remain idle until the operator can replace the tool. If the machine is running unattended (overnight), this could be a very long time.

Since a tool breakage detection system is often used as part of a flexible manufacturing system, it may be desirable to have the machine simply stop working on the current workpiece and go on to the next workpiece in the system. This of course assumes that the same tool is not required for use with the next workpiece and that a tool-life management system is not employed to change to the next tool in the tool group. If a workpiece is only partially machined when the tool breakage detection system causes the balance of machining to be skipped, the system must have some way of alerting the operator that more machining is necessary on this workpiece after tool replacement.

As you can imagine, detecting tool breakage is only part of the problem of utilizing a tool breakage detection system. Determining what to do after the tool breakage is detected (especially with unmanned operation) can be even more complicated.

Conclusion to tool breakage detection systems

Unfortunately, none of the tool breakage detection systems available today is completely fail-safe. None can incorporate the common sense available from even an entry-level CNC operator. They all have their drawbacks and limitations. Even so, they improve the potential for safe operation during unmanned operation.

As time goes on, and as tool breakage detection system manufacturers improve the methods by which tool breakage detection systems work, they will play a much more important role in the usage of CNC equipment.

Adaptive Control Systems

The goal with an adaptive control system is to achieve a consistent metal removal rate with a cutting tool from one workpiece to the next. Since there is a direct correlation between metal removal rate and spindle horsepower, the spindle drive motor as well as axis drive motors will be used to control metal removal rate. As with the first form of tool breakage detection system, an adaptive control system is taught how the cycle should run. With a workpiece of known hardness and known material to be removed, the cutting tool is allowed to machine at the desired cutting rate. The adaptive control monitors and records the horsepower draw during the machining cycle. However, an adaptive control system, instead of simply monitoring for tool breakage during the machining of subsequent workpieces, will actually manipulate cutting conditions (predominantly feed rate) in order to match the model run.

This allows the CNC machining center to adapt to inconsistencies in workpiece material from one workpiece to the next. For example, say a 6-in-diameter face mill is used to machine a large surface of a cast-iron workpiece. Say that there is about 0.150 in stock on the face of the workpiece to be removed. For the first (teaching) workpiece, based on the amount of stock to be removed, say the cutting conditions are set to render a 20 in^3/min metal removal rate.

When the next workpiece is run, say there is only 0.120 in stock on the face of the workpiece to be machined. The adaptive control system will actually increase the feed rate in order to increase the metal removal rate to the desired level.

Keep in mind that it is spindle and axis drive horsepower that is being used to control cutting rate. For this reason, other variations in the machining operation, like material hardness and tool wear, will also affect the system. For example, the adaptive control system will automatically slow the machining operation for harder workpieces. It will also slow the machining operation as the tool dulls, since a dull tool requires more horsepower than a new tool. For these reasons, a well-tuned adaptive control system will achieve maximum cutting conditions (hence maximum productivity) while actually prolonging tool life.

Since adaptive control systems monitor spindle and axis drive horsepower in order to control the machining operation, machining operations that require larger percentages of the machine's available horsepower make the best candidates for adaptive control systems. Adaptive control systems work best with tools like large end mills, face mills, and large drills. For small tools, or for large tools that require small metal removal rates, the adaptive control system will not be able to monitor horsepower changes well enough to optimize cutting conditions without causing tool breakage.

3

Turning-Center
Accessory Devices

Now we will set the focus of our presentations on turning centers. Keep in mind that certain accessory devices are common to both machining centers and turning centers. However, most are applied in a slightly different manner to turning centers. For this reason, while we will assume that you have read the previous machining-center-related presentation for a given accessory device, we will attempt to make each presentation within this chapter as independent as possible.

Work-Holding and Work-Support Devices

How a workpiece is held and supported during its turning-center machining operation is critical to the success of a turning-center operation. If a workpiece is held and supported properly, all other facets of the machining operation will be easy to control. However, an improperly held and supported workpiece will, at best, make the machining operation difficult to perform. In worse cases, an inadequate work-holding and/or support device will make it impossible to hold size on the workpiece. At worst, a poor selection of work- holding and support devices can make for a very dangerous turning-center operation environment.

In this section we will take a close look at the most common work-holding and support devices available for use with CNC turning centers. While we cannot possibly discuss every detail of each device, this presentation should give you a well-rounded view of the devices available. We will begin with work-holding devices and proceed from the most common and easiest to apply to those that are less common and more difficult to work with. Finally, we will discuss work-support devices that are used with CNC turning centers.

Work-holding devices

As you know, a turning center rotates the workpiece as the machining operation takes place. A stationary cutting tool is brought into contact with the rotating workpiece. During any turning operation, the workpiece will be under two kinds of stress.

One cause of workpiece stress is the centrifugal force generated with workpiece rotation. At the high spindle speeds available from today's turning centers, an out-of-round workpiece will have the tendency to shake itself apart.

The second cause of workpiece stress is caused by the machining operation itself. As any cutting tool comes into contact with the workpiece, it will have the tendency to push the workpiece out of the work-holding device. Regardless of the size of the workpiece or the kind of machining operation to be performed, these two stress-related problems *must* be overcome with the work-holding and/or work-support devices used.

As you can imagine, the application for the CNC turning center determines what kind of work-holding and work-support device is required. When they are buying a CNC turning center, most companies attempt to purchase one form of work-holding device and one form of work-support device to accomplish all of their work-holding and work-support needs. Though this may sometimes be possible when there is a small diversity in workpieces to be machined, as the number of different workpiece configurations grows, consideration *must* be given to purchasing more than one work-holding and/or work-support device if safe and efficient machining is to be ensured.

It has been our experience that during the typical purchase of a turning center, too little concern is given to possible work-holding and work-support alternatives. And, as stated, a poorly chosen work-holding and/or work- support device will make for a poor machining environment. In this section, you will be exposed to the most common work-holding alternatives, how they are used, and for what kinds of work they are best.

Try *not* to view a turning center's work-holding device as an integral part of the machine tool itself. Rather, try to view it as only part of the machine's setup, just as a common table vise is only a part of a vertical machining center's setup. Though a table vise is a very common work-holding tool, it will not properly hold all workpieces to be machined by a vertical machining center. Just as a table vise will not suffice for all machining-center setups, no single turning-center work-holding device will suffice for all kinds of workpieces that can be machined by a turning center.

Many companies try to do too much with a turning center's work-holding device. Most have one work-holding device for their turning center, and it has *never* been removed from the turning center. In fact, there are people in this industry who do not even know their turning center's work-holding device can be removed! Please do not misunderstand our point. We are not saying a company should have more than one form of work-holding device if it is not required. We *are* saying that many companies misapply the one form of work-holding device they happen to have without even considering possible alternatives. To make the most efficient use of any turning center, the work-holding device must be wisely chosen for the each workpiece to be machined. The higher the production quantities to be machined, the more important this point.

Three-jaw chucks.   The three-jaw chuck can be thought of as the table vise of turning-center work-holding devices. This form of work-holding device is, by far, the most popular device used to hold turning-center workpieces during machining. Almost every company that utilizes CNC turning centers owns at least one form of three-jaw chuck. Here we will present a cursory view of the types available as well as their basic applications.

Figure 3.1 shows a common three-jaw chuck used with CNC turning centers. Note that this drawing shows no top tooling. We are currently stressing the basic components of the chuck itself.

As you can see, a three-jaw chuck has three moving master jaws. These jaws are serrated in one fashion or another to accommodate the top tooling used to clamp onto the workpiece itself. The master jaws are mounted 120° apart, allowing symmetrical clamping around the workpiece. As the chuck is activated, the three jaws move into

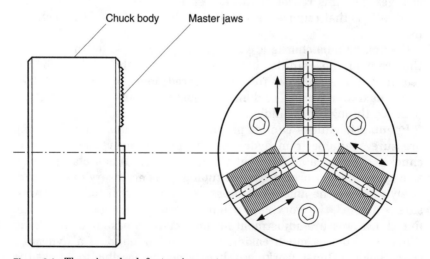

Chuck body   Master jaws

Figure 3.1   Three-jaw chuck for turning-center use.

the clamped position. Note that a three-master-jaw chuck is designed to clamp in both directions, giving it the ability to hold workpieces on outside diameters as well as on inside diameters.

Most three-jaw chucks used with CNC turning centers are activated by a foot switch. By pressing the foot switch once, the jaws close. Pressing it again makes the jaws open. The two most popular styles of three-jaw chuck are hydraulic and pneumatic.

Hydraulic-style chucks require an activator to be mounted at the rear of the spindle. A draw tube (or draw bar) runs through the spindle and moves fore and aft as the chuck is activated to provide the driving force to open and close the chuck. Note that the draw tube, since is within the spindle of the machine itself, reduces the maximum size of bar stock that can be fed through the spindle. In fact, for very small chucks, a draw *bar* is used for activation instead of a draw tube, reducing the usable spindle hole size to nothing and eliminating the possibility of feeding bar through the spindle.

The major advantage of hydraulic chucks is that they provide a tremendous amount of pressure at the jaw, allowing very high chucking pressure to hold the workpiece. This gives the hydraulic-style chuck the ability to securely hold very heavy workpieces.

The major drawback of hydraulic chucks is that they have a very limited range of pressure settings. In fact, it can be very difficult to truly tell exactly how much chucking pressure is at the jaw. Since the chuck pressure is changed by changing the flow of oil into the chuck's activator, precise pressure changes are next to impossible. Also, the amount of pressure supplied by a hydraulic chuck may vary throughout the day, as the temperature of the oil within the activator device changes. For this reason, hydraulic-style chucks are best suited for applications that require a great deal of chucking force, but no chucking finesse.

Pneumatic (air) chucks are activated by air. Most are mounted to the front of the spindle, meaning no draw tube or draw bar is required. Since no draw tube is required, most pneumatic chucks allow maximum workpiece diameters just under the spindle hole size to be fed through the spindle.

Pneumatic chucks cannot provide anywhere near the clamping pressure available from hydraulic chucks. However, their range of chucking pressure adjustment is much better. When clamping on somewhat flimsy workpieces (like tubing), it is necessary to more precisely regulate the amount of pressure at the jaw. If too much pressure is applied, the workpiece will be deformed during chucking. The fine chuck pressure adjustment for pneumatic chucks makes holding flimsy workpieces much easier. On the other hand, if a hydraulic chuck holds a flimsy workpiece, it may not be possible to adjust the chuck pressure precisely enough to adequately hold the workpiece without deforming it.

Top tooling. The type of top tooling best suited to hold the workpiece varies dramatically from job to job. The most common form of top tooling used with three-jaw chucks is called a *jaw*. Here we present the three most common forms of chuck jaws.

Hard jaws. Hard jaws are so named because they are hardened and ground. This hardening process makes the hard jaw harder than the workpiece it is designed to hold. For this reason, hard jaws have the tendency to bite into the workpiece being held.

Figure 3.2 shows one type of hard jaw. Hard jaws are used when it is mandatory to gain the best possible grip on the workpiece. The sacrifice for this powerful grip is that hard jaws may leave severe witness marks on the workpiece. As long as the witness marks will cause no damage to the workpiece, or as long as the surface gripped by the hard jaws will be machined in a later operation, hard jaws make the best possible choice if gripping potential is of primary concern, as would be the case with powerful roughing operations.

Note that most hard jaws are designed to grip on more than one surface of the jaw. As Figure 3.2 shows, most hard jaws can be placed

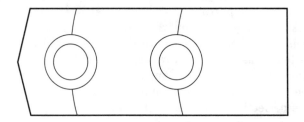

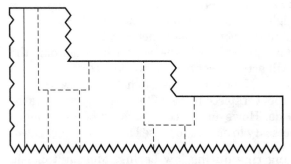

Figure 3.2 Hard jaw for three-jaw chuck. Note serrations that will bite into the workpiece being held.

in the chuck one way to allow relatively small diameters to be gripped, or turned around in the chuck to grip on larger diameters. Keep in mind that this is but one style of hard jaw. Hard jaws can be designed to suit a wide variety of workpiece configurations.

Generally speaking, the biting ability of hard jaws makes them an excellent choice for gripping on rough stock diameters that have not been previously machined. Castings, forgings, and hot-rolled steel are examples of materials that have very rough surfaces in their premachined state, and require the biting ability of the hard jaw.

Keep in mind, however, that hard jaws do not make a very good choice for clamping on finished surfaces for two reasons. First, as mentioned previously, they tend to leave nasty witness marks on the workpiece. These witness marks are usually unacceptable if left on finished surfaces. Second, due the inconsistency of biting depth when hard jaws are used, it is not possible to hold perfect concentricity from the gripped end of the workpiece to the end of the workpiece being machined. If concentricity is required from one end of the workpiece to the other, another form of top jaw must be used.

Soft jaws. Soft jaws are best applied when leaving a witness mark on the workpiece is unacceptable, and/or if concentricity is critical from one end of the workpiece to the other. Unlike hard jaws, soft jaws are not hardened and ground. They are made from steel and left in their soft state. If concentricity is important, soft jaws must be actually machined in the chuck for every setup in which they are used. The diameter of the jaws is machined very close to the diameter of the workpiece that is to be held. Figure 3.3 is a drawing of a soft jaw.

Say, for example, there is a 3-in outside diameter that must be used as the chucking diameter. Say this diameter has been machined in a prior operation, so it is perfectly round and quite smooth. The operator or setup person would mount the soft jaws in the three-jaw chuck and bore the jaws to a diameter very close to a 3-in diameter. Ideally, the soft jaws should be bored just slightly *smaller* than 3 in (say 2.995 in). Once bored, the jaws will have the tendency to expand in diameter slightly. If bored slightly smaller than the chucking diameter, the jaws will nicely expand to the workpiece diameter to be held. If bored too large, the jaws will not properly hold the workpiece, since only the very center of the jaw will actually contact the workpiece.

Tips for boring soft jaws. If there is no need to hold a critical concentricity from one end of the workpiece to the other, just about any fashion of jaw boring will do. However, as concentricity becomes more important, it will be necessary to adhere to these basic jaw boring rules.

1. Clamp on chucking ring during jaw boring. Most automatic three- jaw chucks have a relatively limited jaw stroke (from about

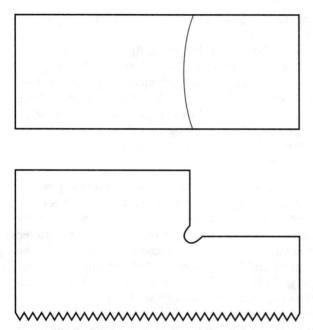

Figure 3.3 Soft jaw for three-jaw chuck. These jaws must be machined as part of the workholding setup.

0.200 to 0.500 in). In order for jaws to actually clamp down on the workpiece after boring, the jaws must be clamping on something during the jaw boring process. If the jaws are simply set to their closed position when jaws are bored, there will be no jaw stroke left to perform the clamping on the workpiece.

While there are special tools designed to be used as chucking rings, most operators use scrap pieces that happen to be lying around the turning center. The chucking ring must, of course, be placed beneath the intended depth of jaw boring to ensure that the jaws can be bored to the desired depth without contacting the ring.

2. Clamp on the chucking ring close to the middle of the jaw stroke. To ensure consistent chucking force from workpiece to workpiece, the operator must clamp on the chucking ring as close to the middle of the jaw stroke as possible. It is at this jaw position that the chuck reaches its best and consistent pressure.

3. Clamp on the chucking ring with the same chuck pressure that will be used to hold the workpiece. When concentricity is highly important, it is mandatory to perform the chuck boring procedure with the chucking conditions set as they will be during the machining of the workpiece. This includes chucking pressure. If chucking pressure is changed after the chuck boring operation, the concentricity of

the bored jaws will change, directly affecting the concentricity of the workpiece.

4. Machine the jaws to the proper diameter. As stated, it is necessary to machine the jaws to a diameter very close to the workpiece chucking diameter. For holding on outside diameters, this means boring the jaws within 0.002 to 0.005 in smaller than the workpiece diameter (the closer to the workpiece diameter, the better the concentricity that can be expected). For chucking on inside diameters, the chuck jaws must be turned to within a close amount *larger* than the holding diameter.

It can be extremely difficult to measure the jaw diameter that has just been machined. Since there are three jaws, and since they are 120° apart, they do not easily lend themselves to being measured by common toolroom gages. In most cases, a micrometer or dial caliper *cannot* be used. Just about the only way to measure a bored diameter is to use a three- pronged dial bore gauge. Unfortunately, most companies do not wish to go to the trouble of purchasing and setting up dial bore gauges for the rudimentary task of boring soft jaws.

For this reason, the operator or setup person may be working blind when boring soft jaws. He or she may not even be able to measure the diameter just machined. How then can an operator or setup person confirm the diameter of the jaws just bored?

You would be surprised at the number of so-called experienced setup people who just "eyeball" the bored diameter. They bore the jaws to a size they think is close. Then they hold the workpiece up close to the bored diameter and try to determine how much more stock is left to be machined on the jaws. They make another pass, and compare again. This is repeated until the setup person is confident the bored diameter is close to the workpiece diameter. Using this technique, the setup person will have no idea as to the size just bored. He or she is simply guessing at the diameter, and there is a good chance the jaws will be bored oversize. Though all of this may sound a little crude (and it is crude!), *many* operators use this poor technique.

In order to precisely bore jaws, the operator or setup person *must* have a way of determining the diameter of the jaws being machined. One way to accomplish this is to utilize the X and Z position displays of the CNC turning center in the same way an operator would use the digital readouts of a manual engine lathe.

To easily control the depth (in Z) of the jaws being bored, the operator can manually bring the tip of the boring bar to the top face of the jaw. Usually a finely graduated handwheel is available and will allow precise control of the machine's movement. Once the boring bar is touching the top face of the jaw, the Z-axis position display can be set to zero. From this point, the Z-axis position display will register the current distance from the face of the jaw to the boring bar tip. As jaws are bored, this makes it very easy for the operator to tell how far into the jaws the boring bar has traversed.

Making this kind of calibration for the diameter (X) axis is a little harder to visualize and accomplish (for the first time), but once done, will allow the operator to monitor the diameter the boring bar will machine by simply looking at the X-axis position display. Keep in mind that this calibration need only be done once for each boring bar used to bore jaws. Since most CNC turning-center operators reserve one or two boring bars for boring jaws, the needed calibrations will be kept to a minimum.

All CNC turning centers have a very accurate position along each axis called the *reference position.* Some control manufacturers call this reference position the *zero return position.* Others call it *grid zero.* Yet others call it the machine's *home position.* No matter what it is called, the reference position is a very accurate location that is commonly used to help with the designation of the program zero point. Additionally, some control manufacturers require that the machine be sent to its reference position as part of the power-up procedure.

For jaw boring purposes, if the operator can determine the diameter of the boring bar tip when the machine is resting at its reference position in the X axis, he or she can easily make the X-axis calibration each time jaw boring is necessary. Here is one way to accomplish this.

First, the operator must chuck-clamp on a piece of tubing. Any workpiece with a hole large enough to let the boring bar enter will suffice. Note that soft jaws do *not* have to be used at this point. Second, the operator will start the spindle and manually bore a small amount of stock from the inside of the hole. The hole must clean up all the way around. Third, without moving the tool in the X axis, the boring bar is retracted from the hole in Z. Fourth, the spindle is stopped and the hole diameter is measured. Fifth, the X-axis display is set to read the diameter of the hole. At this point, the X axis is calibrated with the diameter of the boring bar tip. As the X axis moves, the X-axis display will follow along, constantly showing the current diameter the boring bar will machine. Finally, to ensure the operator's future ability to do the calibration without the need to rebore, the machine is sent to its reference position. At this point, the value of the X-axis position display is documented for the next time chuck boring is required.

For future chuck boring tasks, the operator simply sends the X axis to its reference position and sets the X-axis display to read the previously measured number. From this point, the X axis is calibrated, and the operator can simply monitor the X-axis display to determine what diameter the boring bar will currently machine.

Pie jaws. The tendency of any three-jaw chuck will be to compress and deform the workpiece during clamping. If the workpiece does compress, it will spring back to its normal condition when chuck pressure is released. This will cause any machined surfaces to bend or warp as the chuck pressure is released. The machined diameters will not be truly round. Instead, they will be somewhat egg-shaped. Pie jaws dramatically reduce the tendency for workpiece deformation during chucking. Figure 3.4 shows a drawing of a pie jaw.

As you can see, pie jaws will hold and support the workpiece for almost the entire periphery of the chucking diameter, and tend to keep the workpiece from compressing or deforming during the chucking process. The workpiece will be forced to remain in a round condition from the time it is placed in the chuck.

One limitation of pie jaws has to do with their weight. Since pie jaws extend all the way around the workpiece, they tend to be much

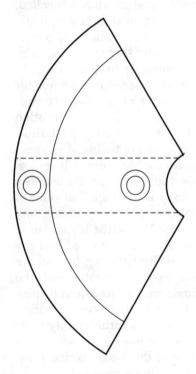

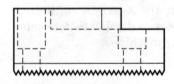

Figure 3.4 Pie jaw for three-jaw chuck. Note that three of these jaws will almost completely surround the periphery of the workpiece.

heavier than normal soft jaws. This added weight can cause problems at high spindle speeds. The faster the spindle rotated, the more centrifugal force will be affecting the pie jaws. This centrifugal force will have a tendency to loosen the pie jaws' grip on external diameters (or increase the grip on internal diameters). For this reason, many companies make pie jaws from light materials, like aluminum.

Programmable features of three-jaw chucks. Almost all three-jaw chucks designed for use with CNC turning centers are automatic, activated by a foot pedal or some form of switch. Additionally, there are other features of three-jaw chucks that are also automatic. Many of these automatic features are often programmable.

Chuck jaws open and close. Of course, the operator must have the ability to manually open and close the chuck jaws in one manner or another. Usually this is accomplished by a foot switch, making it easy to hold the workpiece with both hands during the loading of a workpiece.

Most CNC turning centers also allow the chuck jaws to be opened and closed by programmed commands. Usually two M codes, one for open and another for close, are used for this purpose. The actual M code numbers used for this purpose vary dramatically from one turning center to the next and can be found in the turning center's programming manual.

Though some turning centers allow the jaws to be opened and closed while the spindle is running, most require, for safety reasons, that the spindle be stopped prior to the opening or closing of the chuck. This is true of both manual activation as well as programmed activation.

The most common application for programmable chuck open and close is during bar feeding. Though collet chucks (discussed a little later) make a better choice for bar-feeding applications, a programmable three-jaw chuck will also work. During a bar-feed operation, it is necessary for the bar stock to feed through the center of the spindle. Before this is possible, of course, the chuck must be opened. After the bar is fed, the chuck must be closed before the next workpiece can be machined. Bar feeders will be discussed in detail a little later in this chapter.

Chucking pressure high and low. In most turning-center applications, the workpiece is loaded into the chuck and clamped one time. The chucking pressure (manually selected by a valve or rheostat) will be used to hold the workpiece during the entire machining operation. For a one-time chucking operation, the chuck pressure should be adequate to hold the workpiece rigidly during machining operations, yet *not* be so great as to deform the workpiece during chucking.

With structurally strong workpieces, the operator or setup person will be able to easily make this chucking pressure compromise. However, there will be those weaker workpieces that cannot be completely machined without sufficient clamping force to deform the workpiece, especially if powerful roughing operations must be performed. If only one clamping pressure is used for the entire machining cycle, it may be impossible to successfully machine the workpiece without deformation.

Keep in mind that rough machining operations almost always require more holding force (chuck pressure) than finishing operations. While the high chucking pressure required for roughing operations will often deform the workpiece, the lower chucking pressure required for finishing operations will seldom deform the workpiece.

For structurally weak workpieces, the chucking pressure must be reduced after all roughing operations are performed. Some companies, when faced with this problem, have the operator manually change the chucking pressure during the machining cycle. With the chucking pressure set quite high, the workpiece is loaded and the rough machining operations are performed. After roughing, the cycle is stopped (with a program-stop M00). The operator then lowers the chucking pressure and reactivates the cycle. The finish machining operations are then performed at the lower chucking pressure. This ensures that the workpiece will not be deformed during the finishing operations. When the cycle is completed, the operator *must* remember to increase the chucking pressure *before* the next workpiece is loaded.

Keep in mind that, if a hydraulic chuck is being used to hold the workpiece, it can be *very* difficult for the operator to precisely set the chucking pressure consistently from time to time. In most cases, a chuck pressure gauge must be used to confirm the chuck pressure. This gauge is actually clamped on to give the chuck pressure reading.

Using this method to confirm chuck pressure is tedious and time-consuming. The operator may have to adjust the chuck pressure valve two or three times in order to get the right pressure. This must be done twice during each cycle. And if the operator forgets to change the chucking pressure, at best the workpiece will be scrapped. At worst, the workpiece could be thrown from the chuck.

A programmable high-/low-pressure three-jaw chuck improves this situation. Two M codes, one for high-pressure chucking and another for low-pressure chucking, are used to change the chucking pressure. The operator sets two rheostats (one for high clamping pressure and another for low) during setup. If a chuck pressure gauge is used to confirm chucking pressure, the operator need only use it one time during setup (not twice for each workpiece). In the program, an M code tells the control which chucking pressure to use. Operation becomes much more automatic.

Almost all three-jaw chucks require that the jaws be reclamped during the changing of the high/low valve. This is even true of most programmable high-/low-pressure chucks. During this reclamping process, the workpiece must be held securely in order to maintain concentricity and to keep the workpiece from falling out of the chuck. To allow totally automatic operation, most companies incorporate some form of workpiece pusher that holds the workpiece firmly in place during the reclamping required for changes in chuck pressure.

Chucking direction.   Three-jaw chucks can clamp in both the open and closed position, allowing the chuck to hold on inside as well as outside diameters. For outside diameter clamping, of course, the clamping direction is inward (toward the center of the chuck). For inside diameter clamping, the clamping direction is outward.

Almost all CNC turning centers allow an easy way of changing the clamping direction. Usually a toggle switch or manually activated valve controls which way the chuck clamps. Some even allow the chuck direction to be changed by programmed command.

Note that, for safety reasons, most CNC turning centers will not allow the spindle to come on if the chuck has not been brought to the clamped position, so the chucking direction must be correctly selected before a program can be activated. Also for safety reasons, most turning centers do not allow the chucking direction to be changed while the spindle is running.

Some (but not many) CNC turning centers allow the chucking direction to be commanded within the program. If available, two more M codes are used. One M code selects the inward clamping direction and another selects the outward clamping direction.

There are limited times when the clamping direction must be programmed. Programmable chucking direction is *not* normally required for a CNC turning center if an operator will be loading workpieces. Just about the only time when this feature must be programmed is with certain automation systems. If, for example, an automation system is used to load workpieces, no operator will be available to change the setting of the chuck direction. Say a workpiece must be held in multiple clampings. Say the workpiece must be turned around in the chuck during the cycle and one clamping end requires outside-diameter chucking while the other end requires inside-diameter chucking. In this case, if the operation is to be totally automatic, the chucking direction *must* be programmable.

Collet chucks.   Though not nearly as versatile as three-jaw chucks, collet chucks are commonly used, and provide the CNC turning center user with many benefits. Figure 3.5 shows a drawing of a collet chuck. Instead of incorporating top jaws with which to grip the workpiece, a collet chuck utilizes an internal hardened collet. The outside

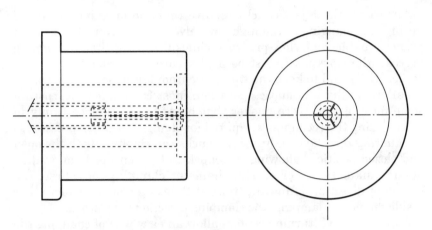

Figure 3.5 Collet chuck for turning-center use.

configuration of all collets used with the collet chuck will remain identical from one collet to the next. Only the inside diameter of the collet (the diameter that holds the workpiece) will change. This diameter is machined just slightly larger than the workpiece diameter to be held. Collets can be purchased as standard items from most manufacturing tooling supply companies and come in standard sizes ranging from 0.0312 to well over 2.0 in in diameter.

The first benefit of a collet chuck is related to the fact that collets are normally available as standard items. Since standard, off-the-shelf collets that are already machined to the correct size can be used, and since it is very easy to exchange collets, work-holding setup time is dramatically lower with collet chucks as compared to three-jaw chucks. Generally, a collet can be changed in less than 2 minutes.

Like pie jaws used on a three-jaw chuck, the collet inside a collet chuck holds the workpiece nearly all the way around the periphery of the workpiece. This gives the collet chuck excellent concentricity qualities.

Since collets are made only slightly larger than the diameter they are designed to hold, they must collapse only about 0.020 to 0.030 in. The time it takes for this small motion is very short compared to the time it takes a three-jaw chuck to complete the closing and opening motion.

Most collet chucks are smooth around their outside diameter. Since they need not hold top jaws that protrude from the face of the chuck, there is no wind resistance during rotation. A three-jaw chuck can sometimes resemble a fan in this regard if long top jaws are used. For this reason, collet chucks take less energy to rotate.

Along these same lines, since the basic clamping components are all internal to the collet chuck, and since the maximum capacity of most collet chucks is under 2 in, collet chucks tend to be quite light. All of this equates to faster spindle acceleration and deceleration time, reduced electricity requirements, and fewer tooling interference problems.

In fact, many CNC turning centers that are designed to be bar machines (used exclusively for bar-feed applications) have the collet chuck as an integral part of the spindle. In essence, these machines have no chuck. The spindle of the machine acts as the collet chuck. As you can imagine, machines designed in this manner will require even less power to drive the spindle than machines with external collet chucks.

Bar feeding with collet chucks. For the reasons given, a collet chuck makes an excellent choice of work-holding device for bar-feed applications. Note that bar feeders will be discussed later in this chapter.

Bar feeding is one form of automation that dramatically reduces the amount of operator intervention required during the machining cycle. A long bar is used as the rough stock. The bar is fed through the spindle and out the collet chuck. A workpiece is machined and cut off. Then another workpiece length of the bar is fed. This process is repeated for the entire length of the bar.

The clamping and unclamping of the collet chuck *must* be a programmable feature if the collet chuck is to work within a bar-feed application. Of course, the collet must release the bar before the bar can be fed.

Another benefit of collet chucks that applies to bar feeding has to do with the configuration of the bar to be held. The bar to be machined can be in the form of round stock, square stock, or hexagonal stock. Though the collet holding the rough stock must be purchased accordingly, a wide range of collet configurations is available. Figure 3.6 shows some of the collet styles.

Using a collet chuck for chucking applications. Though their application in this regard is somewhat limited, certain chucking applications apply nicely to collet chucks. If a workpiece has been cut to length, and if the outside diameter of the rough stock is quite round and smooth (as is ground stock), the collet chuck user can look forward to excellent concentricity qualities when the collet chuck is used.

Unfortunately, collet chucks are not nearly as versatile as three-jaw chucks when it comes to chucking applications. As the complexity of the workpiece in its rough state increases, the potential for applying a collet chuck decreases. If, for example, the workpiece must be supported along the Z axis, a collet chuck cannot be easily used.

Two-jaw chucks. Though the vast majority of workpieces to be machined on a CNC turning center are generally round in their

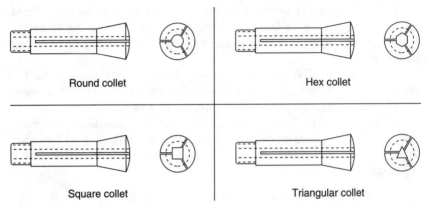

Figure 3.6 Various types of collets that can be used with a collet chuck to grip the workpiece.

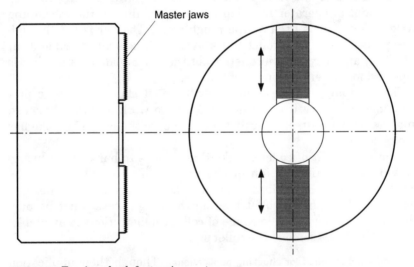

Figure 3.7 Two-jaw chuck for turning-center use.

rough state, not all are. Those workpieces that are not round in their rough state can present quite a work-holding challenge. Figure 3.7 shows a style of chuck that can help hold odd-shaped workpieces, the two-jaw chuck.

A two-jaw chuck is best applied when it is necessary to sandwich the workpiece in the work-holding device. Two jaws mounted 180° apart squeeze the workpiece being held. Figure 3.8 shows an example application for a two-jaw chuck.

Notice that one end of this workpiece has flat surfaces from which the workpiece must be held. The round diameters are the surfaces of the workpiece that must be machined in the turning-center opera-

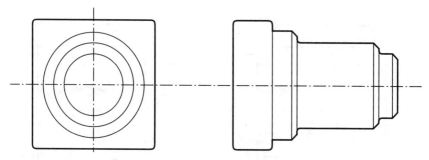

Figure 3.8 Sample application workpiece for two-jaw chuck.

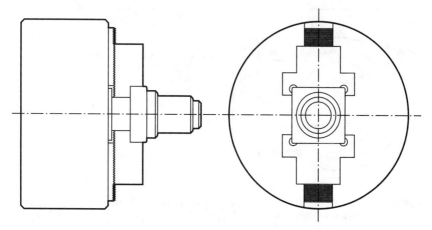

Figure 3.9 Sample workpiece being held by two-jaw chuck. Note that the jaws used with a two-jaw chuck must be specially machined for the workpiece to be held.

tion. Figure 3.9 shows how this workpiece can be held in a two-jaw chuck.

Keep in mind that the soft jaws used to hold the workpiece can no longer be machined within the turning center (as most soft jaws can). These jaws must be specifically designed and made for the workpiece to be held. Also note that many times the workpiece to be held by the two-jaw chuck will not be symmetrical, and the entire setup (chuck, jaws, and workpiece) must then be balanced to allow the workpiece to be rotated at machining speeds.

Index chucks. There are times when it is desirable (if not mandatory) to machine the entire workpiece in one holding. Doing so minimizes workpiece handling and improves the accuracy of the workpiece, since it need not be removed from the setup until it is completely machined. Figure 3.10 shows such an example workpiece that requires machining on more than one side.

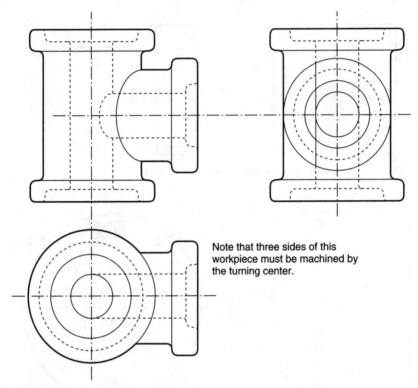

Note that three sides of this workpiece must be machined by the turning center.

Figure 3.10 Sample application workpiece for index chuck. Note how this workpiece must be machined on three sides.

Note that this particular workpiece requires machining on three sides. However, depending on the application, other workpieces may require machining on two or four sides. Some plumbing fittings even require machining on a slight angle (about 5°) from the main centerlines of the workpiece.

An index chuck allows the workpiece to be rotated (swiveled) to allow two or more sides of the workpiece to be exposed for machining. Figure 3.11 shows an index chuck.

The advantages of the index chuck may be self-evident. Since fewer setups are required for each workpiece, there will be less workpiece handling. Since the entire workpiece is machined in one setup, better accuracy can be held from side to side. Since the index is executed by a programmed command, less manual intervention is required.

Unfortunately, index chucks are among the most difficult chucks for which to design and make tooling. Like many jaws made for two-jaw chucks, the jaws for index chucks must be specially designed and made for each workpiece configuration to be machined.

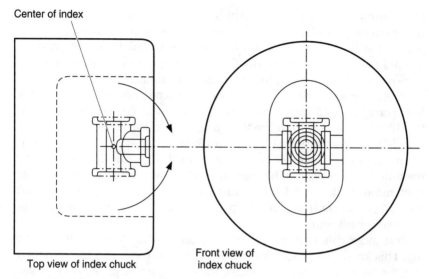

Center of index

Top view of index chuck

Front view of index chuck

Figure 3.11   Index chuck holding sample workpiece.

Due to the complexity of making jaws, index chucks are most often applied when the workpieces to be machined by the turning center are in a family of similar workpieces. Since the clamping jaws for the index chuck must be specially designed and machined for each workpiece in the family, the index chuck does not make a very good general-purpose work-holding device. One common application for index chucks is machining flanges, valves, and fittings for the plumbing industry.

Though their application is somewhat limited, index chucks allow the benefits discussed *and* fast changeover times. In this section, we will primarily address how index chucks are programmed. You will see how good programming techniques can help to minimize setup time.

How an index chuck works.   Like other forms of chucks, index chucks utilize hydraulic or pneumatic pressure to activate the clamping mechanism within the chuck. The clamping motion itself is similar to a two-jaw chuck, since two opposing jaws grip the workpiece. However, many index chucks have only one movable jaw. That is, one jaw remains stationary and the other provides the clamping motion. The clamping is activated manually. The operator will load the workpiece and activate a foot switch to clamp the workpiece between the two jaws.

The index of the workpiece is also activated hydraulically or pneumatically and is a programmable function of the chuck. Most index chuck manufacturers utilize M codes to activate the index motion.

For example, one popular machine-tool builder utilizes an M13 for the purpose of indexing. If the index chuck allows more than one position of index, the M13 can be repeated to index to each position, one M13 per index. Eventually, by repeating the index command, the workpiece will come back to its original position.

Other forms of index chucks have one M code per position of index. For example, M13 may bring the chuck to the 90° position, M14 to the 180° position, M15 to the 270° position, and so on.

Index chucks vary with regard to how far they allow indexing to occur. There are index chucks that can rotate a full 360°, making it possible to machine all the way around the workpiece. However, the vast majority have a limitation in this regard. Many allow up to about 270° of rotation, allowing three sides of the workpiece to be exposed for machining.

Most index chucks allow the index to occur while the spindle is rotating. This keeps the programmer from having to command the spindle to stop for each index and minimizes the cycle time required for indexes. Keep in mind, however, that most index chucks have a maximum spindle speed at which they can safely index. Most cannot index at more than about 1500 RPM. This, of course, means the programmer may have to slow down the spindle before commanding an index.

Programming considerations.   There are two main index chuck programming considerations we will address in this section. One is related to the selection of program zero and the other has to do with utilizing subprogramming techniques.

Program zero selection.   As you know, the wise selection of the program zero point always makes programming easier. You know that all coordinates used in the program will be taken from the program zero point (if you are working in the absolute mode). Many times print dimensions can be used as program coordinates if program zero is chosen wisely. This minimizes the math required for programming.

In the X axis, program zero is *always* the center of the spindle in X. This allows all dimensions going into the program to be easily specified in diameter.

For the Z axis, program zero should also be a logical position along the Z axis that allows coordinates going into the program to match print dimensions. For shaft and chucker applications, the extreme right end of the workpiece being machined makes a good program zero point in Z, since most prints are dimensioned accordingly. With this technique, all Z coordinates used during machining will be negative. Figure 3.12 shows the location of program zero at the extreme right end of the workpiece.

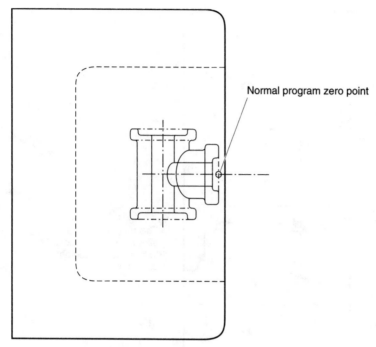

Normal program zero point

Figure 3.12   A common location for the program zero point.

For index chuck applications, however, we do *not* recommend making the extreme right end of the workpiece the program zero point for each side of the workpiece to be machined. The distance from the center of index to the extreme end of the workpiece to be machined will vary from one index position to the next (as Figure 3.13 shows). If the extreme end of the workpiece is used as program zero, after an index all Z coordinates going into the program will have to reflect this difference, making calculations difficult. Also, most prints for parts in this kind of family (plumbing fittings) are not dimensioned from the end of the workpiece. Each end to be machined is usually dimensioned from the center of index.

For this reason, it is wise to make the *center of index* the program zero point in the Z axis. Figure 3.14 illustrates this.

With program zero the center of index, most Z coordinates will be plus (unless machining past the center of index) and will reflect the distance from the center of index to the desired machining position. Also, no special consideration must be given when the workpiece is indexed. Coordinates for all sides of the workpiece will be taken from the same location. Since the drawings for most workpieces to be

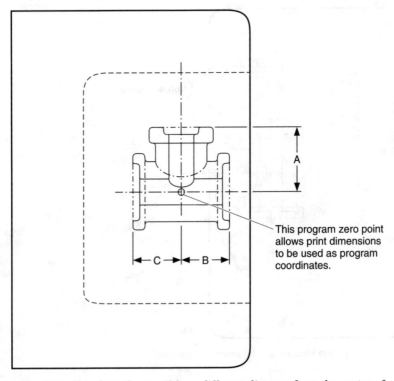

This program zero point allows print dimensions to be used as program coordinates.

Figure 3.13 Note how there will be a different distance from the center of index to the face of each side to be machined. This makes it difficult to use the face of the workpiece as the program zero point.

machined on index chucks are dimensioned from the center of index, print dimensions can be used as program coordinates.

Another advantage of using the center of index as the program zero point has to do with families of parts. Most workpieces that are machined within index chucks fall into a family, such as valves for the plumbing industry. In *many* cases, an entire side of the workpiece is identical from one workpiece to the next. If the center of index is used as the program zero point, it makes it *very* easy to repeat that a certain side be machined from one program to the next. More on this a little later.

Tooling considerations.   Yet another reason to make the center of index the program zero point in the Z axis has to do with the cutting tools. In *all* cases, the location of program zero *must* be specified by one means or another. In most cases, the operator must measure the distance in Z from the program zero point on the workpiece to the tip of each tool while the machine is resting at its reference position.

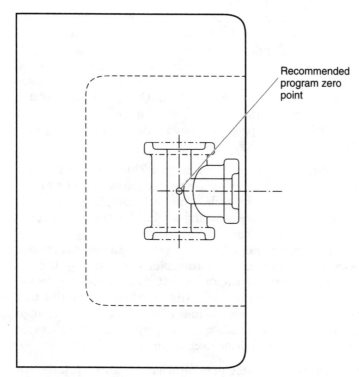

Figure 3.14 The recommended program zero position when using index chucks is the center of index.

This distance is either specified in the program at the beginning of each tool, or within a tool offset.

Keep in mind that some (especially older) CNC turning-center controls have no way of utilizing more than one program zero point within a given program. With these controls, the programmer is *not* free to specify a different program zero point for each side of the workpiece to be machined in the index chuck. With these controls, the only way to effectively program workpieces machined in index chucks is to use the center of index as the program zero point.

Even if specifying more than one program zero point is possible, it means a duplication of effort on the operator's part. He or she will have to duplicate measurements for each side of the workpiece to be machined. The specification of program zero within the program or within offsets will also be duplicated. All of this, of course, means more work for the operator and more time spent setting up. Using the center of index as the program zero point eliminates this duplication of effort.

If the center of index is used as the program zero point from setup to setup, setup time can easily be reduced. Once a measurement is taken from the center of index to the tip of a tool (once per tool), it will remain the same for that tool until the tool is removed from the turret. This, of course, means the operator will not have to take measurements for tools that were used in the previous setup. Understanding this implication is especially helpful when program zero is assigned within the program, but is also helpful when program zero is specified by tool offsets.

When to index. Most companies utilizing index chucks strive to minimize the wear and tear on the chuck by minimizing the number of times they index the chuck per cycle. They will completely machine one side of the workpiece before indexing to machine the next, even if the same tooling is used from one side to the next.

However, in many cases, index time is shorter than the tool change time for chucks that can index while the spindle is rotating. If cycle time is of primary concern, you may have to do some testing to see which occurs faster on the machine, the chuck index or the tool change. If it is substantially faster to index the chuck than to change tools, you may be able to minimize cycle time by machining all sides of the workpiece with one tool prior to changing tools.

Streamlining programs. The last presentation we make relative to index chucks assumes a family of very similar parts. As mentioned, one very common application for index chucks is in the manufacture of plumbing fittings and valves. Many of these parts have somewhat standard ends that determine the size of the fitting. The configuration of one end of a workpiece will be needed often on other parts. For example, one valve to be machined may be required to go from a 2.00-in end fitting to a 3.00-in end fitting. Both these end fittings may be required on other workpieces. If the programs for each end are kept in the CNC control's memory as subprograms, they can easily be invoked whenever required.

For some applications, the various workpieces to be machined require as few as 10 or 12 different programs. If all programs are permanently stored within the CNC control's memory, the programmer or operator can easily execute the needed subprograms from the main program.

As long as all machining is done on one end of the workpiece before an index occurs, subprogramming techniques will drastically reduce programming time. Here is an example program that could be used to call three end machining subprograms. Note that the index chuck used for this example has three index positions. An M13 is used to command the index and is simply repeated to continue indexing to each position. After four M13 commands, the index chuck will be back at its original position.

Main program:

```
O0001 (Program number)
N005 M98 P1000 (Machine 2.00-in end)
N010 M13 (Index to 3.00-in end)
N015 M98 P2000 (Machine 3.00-in end)
N020 M13 (Index to 2.50-in end)
N025 M98 P3000 (Machine 2.50-in end)
N030 M13 (Rotate back to beginning position)
N035 M30 (End of main program)
```

Subprogram for 2.00-in end:

```
O1000 (2-in end)
N005...
  .
  .
  .
(Complete program for 2.00-in end)
  .
  .
  .
N405 M99 (End of subprogram)
```

Subprogram for 3-in end:

```
O2000 (3-in end)
N005...
  .
  .
  .
(Complete program for 3.00-in end)
  .
  .
  .
N350 M99 (End of subprogram)
```

Subprogram for 2.50-in end:

```
O3000 (2.50-in end)
N005...
  .
  .
  .
(Complete program for 2.50-in end)
  .
  .
```

N350 M99 (End of subprogram)

With this technique, the standard programs for each end to be machined can be easily married for any workpiece combination. The relatively simple main program just specifies the order in which the ends should be machined, providing an easy way to come up with the desired workpiece configuration.

If the programmer wishes to machine all sides of the workpiece before changing tools, it becomes a little more difficult to use subprogramming techniques. In this case, each tool of each standard side would require its own subprogram, complicating the situation dramatically.

Work-support devices

Properly holding the workpiece is sometimes only part of a good turning-center setup. In many cases, the work-holding device by itself will not be sufficient to keep the workpiece stable during the machining operation. It is often necessary to supplement the work-holding device with some kind of work-support device.

Generally speaking, the longer the workpiece is, the more probable it is that the work-holding device by itself cannot properly hold the workpiece. To help determine whether some kind of work-support device is required, there is a basic machining practice rule of thumb involving the length-to-diameter ratio: if the workpiece extends from the work-holding device over about 3 times its diameter, and if optimum cutting conditions must be achieved, some form of work-support device will be necessary.

For example, if a 1-in-diameter workpiece is to be machined, it should not extend more than about 3 in from the work-holding device. If it must, some form of work-support device will be necessary.

Tailstocks. The most common form of work-support device is the tailstock. A tailstock is used to support the end of the workpiece opposite the work-holding device. In essence, when a tailstock is used, both ends of the workpiece will be supported. Figure 3.15 shows a setup incorporating a tailstock.

Keep in mind that a tailstock is not equipped on every CNC turning center. If you do not expect to machine lengthy workpieces, you may elect to save some money and *not* purchase the tailstock option.

For CNC turning-center application, a tailstock is made up of three primary components: the body, the quill, and the center.

The tailstock body. The body of the tailstock provides the tailstock with its rigidity. All CNC turning centers that incorporate a tailstock

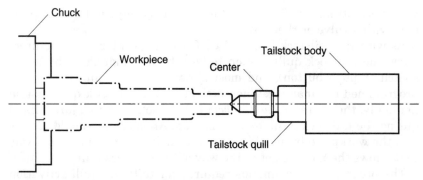

Figure 3.15   Tailstock for turning-center application.

allow the tailstock's body to move to allow for different lengths of workpieces. Almost all allow the body to be moved manually, by the operator, as well as automatically by programmed commands.

Manually, the operator can activate a push button, foot switch, or lever in order to make the tailstock body move. The operator will move the tailstock body into position as part of the workpiece setup.

Machine-tool builders vary dramatically with regard to how the tailstock body is commanded to move within a program. Most utilize two movable cams (called *dogs*) in conjunction with a limit switch mounted on the tailstock body. As part of the setup, the operator moves the dogs to the desired positions. With this arrangement, two M codes are used to command tailstock body movement. One M code tells the control to move the tailstock body forward until the dog is contacted. Another tells the control to move the tailstock body in the reverse direction until the other dog is contacted.

Another popular way machine-tool builders use to move the tailstock is to drag it along with the machine's Z-axis motion. The turret is commanded to move to a locking position. At this position, an M code activates the locking device, which temporarily connects the turret to the tailstock body. The Z axis is then told to move into the desired position. When in position, another M code tells the control to disconnect the tailstock body from the turret.

Spindle considerations.   For safety reasons, most CNC turning-center manufacturers disable the tailstock body motion if the spindle is running. If a command is given to move the tailstock body while the machine's spindle is running, most controls will generate an alarm.

Tailstock quill.   The tailstock quill is the device that actually applies the pressure to the workpiece. Like the tailstock body, it can move forward to contact the workpiece and back to retract from the workpiece. The quill pressure is usually adjustable to allow for different workpiece strengths. The operator can easily adjust the quill pressure with a valve or rheostat.

As with the tailstock body, all CNC turning-center manufacturers allow the tailstock quill motion to be activated manually (by a foot switch or push button), and most allow automatic quill motion (by programmed commands). When activating the tailstock quill motion manually, the operator *must be very careful* so as not to injure a hand during the tailstock quill forward motion. An operator should *always* grip the workpiece from below and *never* hold the workpiece in a way that allows the hand to get in the way of the tailstock quill.

The programmed commands required for tailstock quill activation vary from one turning-center manufacturer to the next. Most use two M codes. One M code brings the quill forward until it contacts the workpiece. Another retracts the quill. The actual quill stroke length also varies from one turning center to the next. And, as with the tailstock body, most turning-center manufacturers disable tailstock quill motion while the spindle is running.

Center.   The center is the device that actually comes into contact with the workpiece. It is removable, allowing for a wide variety of work-support configurations. For small workpieces, it is most commonly in the form of a small pointed cone. The angle of this cone, 60°, must match the center-drilled hole in the workpiece.

Keep in mind that there *must* be some prior machining operation on the workpiece before the center is engaged. Most commonly, a center-drilled hole is used for this purpose. However, larger holes in the center of the workpiece can also be used.

The portion of the center that actually contacts the workpiece is usually allowed to rotate. This form of center is called a *live center.* Another form, one that does not rotate (usually more common on manual engine lathes), is called a *dead center.* The rotation capability is sometimes designed into the center itself, but more commonly is designed into the tailstock's quill.

Alignment problems.   During machining, the center *must* be perfectly aligned with the spindle. If it is not, the tailstock, when engaged, will have the tendency to actually bend the workpiece and cause a taper to be machined on all diameters.

During setup, the operator must check the alignment of the tailstock for taper. After the work-holding setup is made, and after the tailstock has been properly positioned, a test cut is made (manually) on the workpiece. This test cut is made with the tailstock center engaged with the workpiece. The operator will skim-cut a relatively long length of the workpiece and measure the diameters at both ends. If the two diameters are precisely the same, the tailstock is properly aligned. If not, the tailstock is out of alignment and must be adjusted.

How a tailstock is adjusted varies from one manufacturer to another, as does the difficulty involved. Most incorporate a series of locking bolts which must be released. At this point, a set screw is usually turned to physically move the tailstock center position. If a dial indicator is attached to the turret and made to contact the tailstock center close to the workpiece, the operator can easily monitor the amount of motion at the tailstock center.

Unfortunately, this tends not to be an exact procedure. In most cases, the operator must use trial-and-error techniques, and test-cut after each adjustment.

Programming considerations. Though the tailstock body and quill are programmable on most CNC turning centers, there are relatively few applications when the programmability of the tailstock is mandatory. In most applications, the operator will simply activate the tailstock manually as part of the workpiece loading process. For example, if loading a long shaft (with a previously machined center-drilled hole), the operator will (1) place the workpiece in the jaws of the chuck and clamp the chuck, (2) activate the tailstock quill (usually by a foot switch) to engage the quill into the workpiece, and (3) release the chuck jaws and reclamp them to ensure a proper seating of the tailstock center.

One time when it will be necessary to engage the tailstock quill during the program is when the workpiece has not been previously machined with a center-drilled hole. The programmer may wish to center-drill the workpiece within the turning-center operation *before* the tailstock is engaged for the balance of the machining. Here is an example program that will accomplish this. For this particular example, here are the M codes that are being used to activate the tailstock body and quill. Keep in mind that they will change from one machine to the next.

M28 Tailstock body forward

M29 Tailstock body back

M16 Tailstock quill forward

M17 Tailstock quill back

Program:

```
O0001 (Program number)
(Number 5 center drill)
N005 T0101 (Rotate turret to center drill)
N010 G97 S500 M03 (Turn spindle on CW at 500 RPM)
N015 G00 X0 Z.2 (Rapid center drill up to workpiece)
N020 G01 Z0.25 F.005 (Center-drill hole)
N025 G00 Z.2 (Rapid out of hole)
N030 X8. Z6. T0100 (Rapid to tool-change position, cancel offset)
N033 M05 (Stop spindle to engage tailstock)
```

N035 M28 (Bring tailstock body forward)
N040 M16 (Bring tailstock quill forward to engage quill)
N045 T0202 (Index to next tool)

.

.

.

N545 G00 X8. Z6. T0500 (Rapid last tool to tool-change position, cancel offset)
N550 M05 (Stop spindle to disengage tailstock)
N555 M17 (Retract tailstock quill)
N560 M29 (Retract tailstock body)
N565 M30 (End of program)

As you can see, this program automatically brings the tailstock body and quill into position after the center-drilling operation. However, there are some limitations and pitfalls to using this technique. Most experienced machinists would agree that, if the tailstock is required for support, it will be needed for *all* machining operations. With this example there is no end support during the center-drilling operation. The workpiece will be rotating at 500 RPM (in our example program) during the center drilling. If the workpiece is not perfectly centered in the jaws, it is likely that the workpiece may be thrown from the chuck.

Another related problem has to do with whether the center-drilled hole will really be in the center of the workpiece. If the workpiece is not perfectly centered in the chuck, the center-drilled hole will not be properly machined.

Whenever possible, it is *always* better to perform the center-drilling operation *prior* to the CNC turning-center operation if tailstock support is required. In fact, there are special manual machine tools (called *facing and centering machines*) that are specifically designed with the express purpose of facing and center-drilling long shafts.

Another time it is truly necessary to have a programmable tailstock is in bar-feed applications. If, for example, a long, skinny shaft is to be machined, the programmer may elect to center-drill the workpiece just after the previous workpiece has been cut off. Then the bar feed will occur. After bar feeding, the tailstock can be engaged for the balance of the machining operation. When finished, the tailstock can be retracted by the program just before the workpiece is cut off.

Steady rests.   While a tailstock helps immensely with the stability of a workpiece during machining, there are times when even tailstock support is not good enough. As you know, when a tailstock is used, the workpiece will be supported at both ends, and machining at either end will be quite stable.

However, if the workpiece is quite long, and if machining is to occur over the entire length of the workpiece, the workpiece may not be so stable in the middle. It may have the tendency to push away

Figure 3.16 Photograph of steady rest used with CNC turning centers. (*Courtesy Mori Seike USA.*)

from the cutting tool. If this occurs, diameter dimensions in the middle of the workpiece will be difficult to hold. A steady rest can be used to provide additional support. Figure 3.16 shows a photograph of a steady rest and Fig. 3.17 shows a more detailed drawing of how it is used to hold the workpiece.

As you can see, this form of steady rest incorporates three rollers which come into contact with the workpiece. This action provides the needed support for machining close to the middle of the workpiece. Note that two or more steady rests could be placed in tandem if a tool must turn from one end of the workpiece to the other. For this application, one steady rest can remain open while the tool passes through. Once the tool has passed the steady rest, it can be closed and the one in close proximity can be opened.

Most turning centers that incorporate this form of steady rest allow the opening and closing of the rollers to be done by programmed command. Usually two M codes are used for this purpose, one to open and another to close. Some turning centers even allow the positioning of the steady rest along the Z axis to be activated by programmed command.

Another popular application for a steady rest is in machining the very end of a long workpiece. If only external work (turning) is to be done, of course a tailstock can be easily used to accomplish the neces-

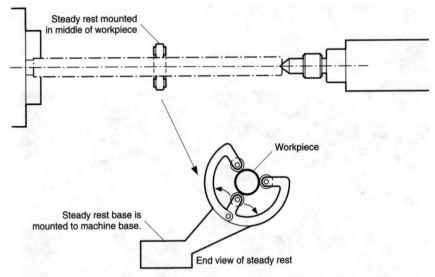

Figure 3.17 Drawing of steady rest showing roller bearings that support workpiece.

sary support. However, if internal work (boring) is to be done, a tailstock cannot be used.

In this case, a steady rest can be used to support the workpiece in a way that allows machining inside the end of the workpiece. It will locate from a previously machined outside diameter and provide the needed support while the inside diameters are machined.

Bar Feeders

Bar feeders are the simplest form of automation systems for turning centers. They feed bar stock through the spindle to provide the rough stock for the workpiece to be machined. For each cycle, a workpiece is machined and cut off. The bar is fed again and another workpiece is machined. This process is repeated for the entire length of the bar. Some bar feeders even have automatic bar-loading systems so that the operator does not have to load bars into the bar feeder. All this, of course, means unmanned operation while the machine runs all the workpieces from the bar. Figure 3.18 shows a common type of bar feeder.

How the bar feeder works

Generally speaking, bar feeders can easily accommodate bars from under 0.25 in in diameter to over 2 in in diameter. Depending on the style of bar feeder, there may be several bushings that are mounted

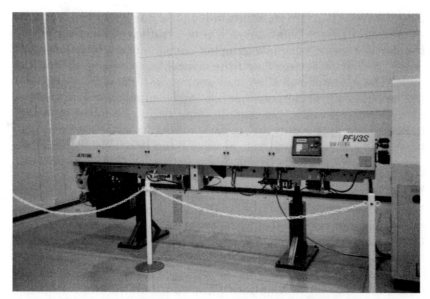

Figure 3.18 Bar feeder attached to a CNC turning center. (*Courtesy Mori Seike USA.*)

within the bar feeder that guide the bar through the middle of the spindle. The hole in each bushing (if required) is machined just larger than the outside diameter of the bar.

Most bar feeders require that the bar be quite straight, and free from bends or bows. You can imagine trying to rotate a crooked 2-in-diameter, 12- ft-long bar at fast spindle speeds. Vibration would make machining next to impossible. Shops that do a great deal of bar feeding have bar straighteners capable of removing bends from long bars.

Some bar feeders require that the end of the workpiece being fed into the machine be chamfered. This allows the bar to easily find its way through the spindle. Those bar feeders that require chamfering usually come with a chamfering device that makes it easy and convenient for the operator to machine the chamfer. Unfortunately, once the bar is fed through the spindle, the chamfer on its end may need to be removed before machining can begin.

Most bar feeders apply a constant feeding force to the bar. As soon as the bar feeder is turned on, it will try to push the bar in the direction of the spindle. As the collet (or chuck jaws) are opened, the bar will immediately feed through the spindle. For this reason, a bar stop mounted in the turret will be programmed to provide the proper stopping position for the bar. More on how a little later.

The bar feeder *must* be interfaced with the machine tool for the purpose of providing an end-of-bar signal to the CNC control. As stated, the machine will continue to machine workpieces for the entire length of the bar. If left unattended, the bar will eventually be exhausted. At some point, there will be insufficient material left on the bar to provide adequate gripping support in the collet or chuck jaws. The bar feeder must tell the control at this point to stop the cycle. The operator will then load another bar, or, if the bar feeder has an automatic loader, another bar will be automatically loaded.

In almost all cases, the spindle must be in a stopped condition before bar feeding can occur. In fact, most turning-center manufacturers will keep the chuck from opening if the spindle is running.

Work-holding considerations

While a three-jaw chuck will work and is sometimes used, most companies that are serious about bar feeding will equip the turning center with a collet chuck. The collet chuck makes the best choice for bar feeding for several reasons.

First, it can support the bar nicely. As compared to other chuck styles, the collet chuck maintains a grip on almost the entire diameter being held for a length of at least 2 in (depending on the collet size). This assures that the rough stock will run true during machining.

Second, collet chucks are light in weight, allowing fast response to spindle speed changes. A collet chuck will take only a fraction of the time a three-jaw chuck takes to respond to spindle speed changes.

Third, all moving and rotating mechanisms are internal to the collet chuck. With any chuck that requires top tooling, like three-jaw chucks, the jaws are mounted on the face of the chuck and act as a fan. The wind created by the rotating jaws can be detrimental to the machining operation (if coolant flow is deflected), and the spindle must overcome the wind resistance as well as the force of the machining operation.

Forth, jaw stroke within a collet chuck is kept to a minimum. Typically, only about 0.030 in (or less) of movement takes place. This means a collet chuck can be opened and closed very quickly. Compare this with a typical three-jaw chuck that may have as much as 0.375 in jaw movement.

Fifth, collet chucks are usually very small in diameter, minimizing any interference problems with tools held in the turret. Therefore, no special consideration need be given to the tool locations in the turret.

Sixth, for turning centers that are specifically designed for bar-feed applications, the collet chuck is actually an integral part of the machine's spindle. It does not extend past the spindle nose. This eliminates *any* possibility for tooling interference problems, makes the setup as rigid as it can possibly be, and allows ultrafast spindle response to speed changes.

Styles of bar feeders

There are two basic types of bar feeders. The difference between these two types has to do with how the bar is supported within the bar feeder itself. One form of bar feeder supports the bar within a series of bushings. The bushings are placed at even intervals for the length of the bar feeder. As the bar rotates, the bushings provide a bearing surface for the bar and keep the bar running true.

This inexpensive form of bar feeder works nicely for relatively small diameter, *straight* bars. For small-diameter bars, the bushings within the bar feeder can easily provide enough support. However, as the bars grow in diameter, and/or if the bars are not perfectly straight, this form of bar feeder is *very* prone to vibration. When these bar feeders vibrate, a great deal of noise is generated.

For this reason, bushing-style bar feeders often require that the turning center be limited with regard to its maximum RPM. For example, once the setup is made, the operator can test the setup to determine how fast the spindle can run without vibration. With a new bar held in the collet chuck, the operator will start the spindle at a relatively low RPM. In small increments, the operator will slowly increase the spindle speed until vibration occurs. Say, for example, vibration occurs at about 1800 RPM. In this case, the program should be limited to something a little less than 1800 RPM to ensure a safety margin. More on how this is programmed a little later.

A better (and more expensive) form of bar feeder is the hydraulic bar feeder. With this style of bar feeder, the bar is supported within the bar feeder in a cushion of oil, instead of by solid bushings. The cushion of oil provides a nice dampening effect, minimizing vibration and noise, even for large-diameter and somewhat crooked bars.

A hydraulic-style bar feeder can rotate at *much* higher speeds without vibration than a bushing-style bar feeder. Though this speed improvement varies with the quality of the bar feeder, most hydraulic-style bar feeders can outperform bushing-style bar feeders by a margin of at least 3 to 1 when it comes to maximum spindle speeds. For this reason, any company that is truly serious about bar feeding should purchase the hydraulic-style bar feeder. The difference in cost can be easily justified by the higher speeds (faster cycle times) they allow.

How to program for bar feeders

As stated, most bar feeders will constantly apply pressure to the bar in the direction of the spindle. For this type of bar feeder, there is no actual command to activate the bar feeder itself. The entire bar-feeding routine is programmed solely by synchronizing the opening and closing of the chuck jaws with the positioning of a stock stop in close proximity to the bar end. As the chuck opens, the bar feeds out. When the stock stop is properly positioned and the chuck is closed, the bar stops moving out. More on this series of movements a little later.

Some bar feeders allow the programmer to turn on and off the pushing pressure of the bar feeder by programmed commands. This minimizes the amount of pressure on the bar during machining. For bar feeders that allow this, usually two M codes are used to control the function. One M code (commonly an M13) activates the pushing motion and another stops the pushing pressure.

Note that some bar feeders even work with a time delay. For these bar feeders, only one M code is required (to activate the pushing). After a certain amount of time, the pushing pressure will cease. The amount of time delay is usually an adjustable feature of the bar feeder.

Determining how much to feed the bar.   Before you can write the bar-feeding portion of your program, you must determine how much the bar must feed. Figure 3.19 shows how to calculate the amount of bar feed to take place. As you can see, you simply add the workpiece length plus the cutoff tool width plus any facing stock to be machined from the workpiece in order to come up with the bar-feed amount.

Some collet chucks tend to suck the bar back into the collet as the collet is closed. If this happens, the length of bar-feed pullout must be adjusted accordingly. Fortunately, this pullback amount will remain fairly consistent from one setup to the next.

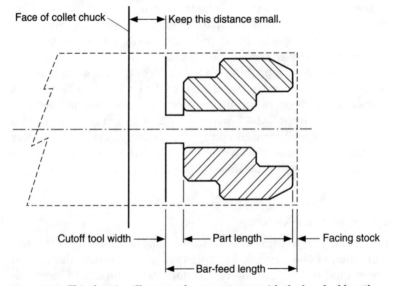

Figure 3.19 This drawing illustrates how to come up with the bar-feed length.

Notice from Fig. 3.19 that the workpiece should be kept very close to the face of the collet chuck. This minimizes the amount of overhang required for machining, which in turn minimizes the potential for vibration and maximizes the potential for machining with optimum cutting conditions. Most companies will try to maintain a constant and safe distance here. About 0.25 in will usually do nicely.

The steps to bar feeding. As stated earlier, the bar feeder will itself apply a constant pressure to the bar, pushing it toward the spindle. It does *not* control how far the bar is fed during each cycle. As soon as the chuck jaws open, the bar will be pushed out.

When bar feeding, some form of stock stop mounted in a turret station must be used to provide a stopping surface for the bar end. During the bar feed, this stock stop will actually move by the amount of the bar feed, in effect, drawing the bar out from the spindle. Some companies use a special stock stop mounted in an individual turret station to stop the bar. However, most programmers consider this to be a waste of a turret station. The shank end of the cutoff tool makes an excellent stock stop and eliminates the need to use a turret station just for the stock stop. Since the cutoff tool will, of course, be very close to the workpiece after cutting it off, it is in an excellent position to provide the stock-stopping function. This also eliminates the cycle time needed to position and index the turret. With these basic points understood, let's look at the four basic steps to bar feeding. Figure 3.20 shows the motions.

1. *Position the tool close to the bar end.* As the upper left drawing in Fig. 3.20 shows, the stock stop is brought to a position very close to the end of the bar. This position must be close enough to the bar end that it does not allow the bar to slam into the stock stop when the chuck is opened. For larger bar diameters, there will be a great deal of inertia built up as the bar moves forward. Most programmers stay about 0.100 in away from the current bar end to keep the bar from building up too much force. Notice that if the cutoff tool is used as the stock stop, the tool must be brought down in X to a position that allows the shank of the tool to be the stock stop.

2. *Open the chuck to allow the bar to come out.* The chuck is opened (by an M code) to release the bar (as the second drawing in Fig. 3.20 shows). If the bar feeder is one that constantly applies pushing pressure, the bar will immediately come out and contact the stock stop. If the bar feeder is one that allows pushing pressure to be turned off, the proper M code to turn pressure on must also be commanded. At the end of this step, the bar will be resting flush with the stock stop.

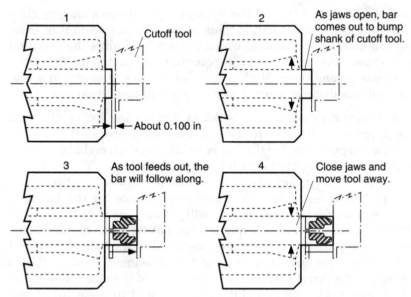

Figure 3.20 The four steps to the bar-feeding process.

3. *Draw the bar out.* As shown in the third drawing of Fig. 3.20, the turret is moved away from the chuck by the feed-out distance calculated earlier. This motion *cannot* be given at the machine's rapid rate. The rate of feed motion in this command must be just slightly *slower* than the bar feeder's traverse rate to ensure that the stock stop will maintain constant contact with the bar. If the stock stop gets ahead of the bar end, the bar will slam into the stock stop at the end of its motion, possibly causing damage to the machine. Most programmers use a feed rate of about 30 to 40 IPM for this motion.

Keep in mind that the spindle is stopped at this point, therefore the feed rate of the pull-out motion *cannot* be given in inches per revolution (IPR). This feed-rate motion must be given in inches per minute (IPM). After the bar feed, the programmer *must* remember to reselect the IPR mode if machining feed rates are to be given in inches per revolution.

A G code is usually used to select the feed-rate mode. One popular control, for example, uses a G98 to select IPM mode and a G99 to select IPR mode.

4. *Close the chuck and move the stock stop away.* As the fourth drawing in Fig. 3.20 shows, the chuck is closed to stop the bar from further motion. The stock stop is then moved away from the workpiece (first in the plus *Z* direction) and the bar-feeding operation is completed.

The redundancy of bar-feed programming. As you can imagine, the series of commands required for bar feeding will be very similar from one setup to the next. As long as the cutoff point remains the same from workpiece to workpiece (usually about 0.250 in from the collet chuck face), only one command in the entire bar feed routine will change—that giving the pullout distance the stock stop must move during the bar feed itself.

Since bar-feeding commands are so similar from one setup to the next, most companies that bar-feed often use a subprogram in which to store the bar-feeding routine. This subprogram remains in memory from one setup to the next, and the operator simply changes the pullout amount as part of each new job's setup.

If the CNC control allows parametric programming techniques, programming the bar-feed operation can become even simpler. Once the parametric bar-feeding program is written, the programmer can simply call the parametric program from the machining program, passing a variable equal to the bar pullout amount for use during the execution of the parametric program.

When to program the bar feed. Some companies like to program the bar-feed cycle at the very beginning of the program, before any machining takes place. Others like to program it at the end of the program, after the cutoff operation. By and large, this is just a matter of personal preference. However, the method chosen in this regard has a great deal to do with how the operator will position the bar as it is loaded into the spindle at the very beginning of the bar.

If the programmer elects to bar-feed at the very beginning of the program, the operator will load the bar into the spindle with the bar end the same distance from the collet chuck face as it will be just after cutoff. When the cycle is activated, the bar feed will occur in the normal manner. With this method, the operator need not position the bar perfectly. If anything, the operator will want to position the bar slightly on the short side to ensure that the stock stop will have sufficient clearance.

If the programmer elects to bar-feed at the end of the program, the operator must load the bar into the spindle with the bar end extending to the same position as after the bar feed (ready for the first workpiece to be machined). In this case, it is a little more critical that the operator position the bar properly. If it extends out of the chuck too far, the first tool will have more stock to machine from the face of the workpiece. If it does not extend from the chuck far enough, the face of the workpiece may not clean up.

Ending a bar-feed program. A bar-feed program must be totally automatic. Therefore, at the end of the machining program, the control must not stop. It must return to the program's beginning and execute the program again for the next workpiece. And, of course, this process must be repeated for the entire length of the bar. The CNC program must be executed as many times as there are workpieces in the bar.

CNC control manufacturers vary with regard to how the program is repeated. One popular control uses an M99 to end the bar-feeding program. The M99 tells the control to return to the beginning of the program and continue running without hesitation. In essence, the control is put into an endless loop. It is not until the control receives the end-of-bar signal from the bar feeder that the cycle is halted.

Other control manufacturers use an on/off switch mounted on the control panel to control automatic operation. How this switch is labeled varies. Some manufacturers call it *continuous operation.* Others call it *automatic operation.* No matter what this switch is called, when it is on, it tells the control not to stop at the execution of the program ending work (usually M02 or M30). Instead, the control will return to the beginning of the program and continue with the next workpiece. This process is repeated until the control receives the end-of-bar signal from the bar feeder.

One last method we will mention has to do with bar feeders that are not fully interfaced with the CNC control. Note that these bar feeders can be dangerous to operate. They do not have an end-of-bar signal, so the control will have absolutely no way of determining when the workpiece being machined does not have an adequate chucking length. Left to continue machining, this form of bar feeder will never stop on its own. As stated, these machines can be very dangerous to operate. If a workpiece is not properly held in the collet chuck, it could be thrown from the chuck. One way to attempt safe utilization of this style of bar feeder is to determine how many workpieces *can* be safely run from each bar and use subprogramming techniques to machine this exact number of workpieces.

Here's an example. Say you determine that you can safely run 35 workpieces from each bar. Here are two programs that will accomplish this and then stop.

```
O0001 (Main program)
N005 M98 P1000 L35 (Run program number O1000 35 times)
N010 M30 (End of program, shut machine down)


O1000 (Subprogram that runs workpiece)
(Bar-feed and run entire workpiece)
.
.
.
```

M99 (Return to main program)

In line N005 of the main program, the control is told to execute program number O1000 thirty-five times, which in turn runs thirty-five workpieces. Then the machine will stop due to the M30 in line N010 of the main program. This program will work just fine as long as the operator does *not* stop the machine in the middle of the bar. But, as you can imagine, if the cycle becomes out of sync for *any* reason (stoppage for lunch, breaks, or end of shift), the machine will continue running beyond its safe limits.

An example bar-feeding program

Figure 3.21 shows the workpiece to be used for our example. For the purpose of this example, we'll say that M14 closes the chuck and M15 opens the chuck. We'll also say that the bar feeder for this example applies constant pushing pressure so no special command is required to turn on the bar feeder's pushing pressure. Also, this bar feeder utilizes an end-of-bar signal with which to stop the cycle at the end of the bar. Note that our example shows the bar-feeding routine at the beginning of the program; the operator would load the bar with the bar end close to the collet chuck and ready for the bar feed. Only the bar-feeding portion of the program is shown. For the purpose of this example, notice that program zero is set as the very front of the finished workpiece.

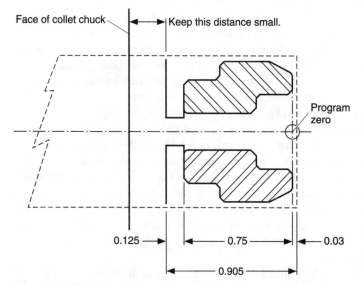

Figure 3.21 Workpiece for bar-feed example program.

Program:

```
O0001 (Program number)
/N001 G00 X5. Z5. (Move to a clearance position to load a new bar)
/N002 M00 (Stop to allow operator to remove remnant and load new bar)
N005 T0505 [Index to cutoff tool station (if it is not already in position)]
N010 G00 X-1.5 Z-.775 M05 (Rapid within 0.1 in of bar, ensure that spindle is off)
N015 M15 (Open jaws, bar comes out to stop)
N020 G98 G01 Z.03 F35. (Select IPM mode, fast feed to stop position)
N025 G99 M14 (Reselect IPR mode, close jaws)
N030 G00 Z.4 (Rapid away in Z)
N035 X6. Z4. (Rapid to tool change position)
N040...(Machine entire workpiece)
        .
        .
        .
N220 T0505 (Index to cutoff tool)
N230 G96 S300 M03 (Start spindle CW at 300 SFM)
N235 G00 X1.2 Z-.875 (Rapid into cutoff position)
N240 G01 X0.187 F.005 (Cut off to hole)
N245 M05 (Turn off spindle for bar feed)
N245 M99 (End of program, go back to beginning and continue)
```

Note that this example used the absolute mode for all motions. However, if bar feeding is done often, it may sometimes be easier to use the incremental mode to specify the stock stop motions.

Notice that lines N001 and N002 are under the influence of an optional block skip word (the slash code). If the operator wishes to load a new bar, he or she will turn *off* the optional block skip function *before* running the cycle. In this case, the stock stop will move to a clearance position and stop (due to the M00). At this point the operator will remove the remnant of the last bar from the spindle and load a new bar.

When the new bar is loaded, it will be kept very close to the face of the collet chuck. The operator will then turn *on* the optional block skip function so that these two commands will not be executed again for the balance of the bar. When the cycle is activated, the machine will continue to run workpieces until the end-of-bar signal is reached.

At the end of this program, notice that the stock stop (shank of cutoff tool) is left close to the workpiece at the end of this program to minimize the amount of travel required during bar feeding. As the program continues from the beginning for the next bar feed, the stock stop will be close to the workpiece. Only if the optional block skip switch is turned off will the stock stop rapid away from the workpiece.

Also note that this program ends with an M99 command. This tells the control to return to the beginning of the program and continue without stopping.

Keep in mind that this is but an example of how bar feeding can be accomplished. Each company and even each programmer will tend to develop the bar-feed routine in a way that makes the most sense to them.

Bar-feed subprogram example. As mentioned earlier, the bar-feeding routine will be quite similar from one workpiece to the next. For this reason, many programmers elect to make the entire bar-feed routine in the form of a subprogram. Here is an example using the same workpiece shown in Fig. 3.21.

Main program:

```
O0001 (Program number)
N005 M98 P1000 (Jump to bar-feed subprogram)
N010...(Machine entire workpiece)
.
.
.
N220 T0505 (Index to cutoff tool)
N230 G96 S300 M03 (Start spindle CW at 300 SFM)
N235 G00 X1.2 Z-.875 (Rapid into cutoff position)
N240 G01 X0.187 F.005 (Cut off to hole)
N245 M05 (Turn off spindle for bar feed)
N250 M99 (End of program; go back to beginning and continue)
```

Subprogram:

```
O1000 (Program number)
/N005 G00 X5. Z5. (Move to a clearance position to load a new bar)
/N010 M00 (Stop to allow operator to remove remnant and loads new bar)
N013 T0505 [Index to cutoff tool station (if it is not already in position)]
N015 G00 X-1.5 Z-.775 M05 (Rapid within 0.1 in of bar, ensure that spindle is off)
N020 M15 (Open jaws, bar comes out to stop)
N025 G98 G01 Z.03 F35. (Select IPM mode, fast feed to stop position)
N030 G99 M14 (Reselect IPR mode, close jaws)
N035 G00 Z.4 (Rapid away in Z)
N040 X6. Z4. (Rapid to tool-change position)
N045 M99 (Return to main program)
```

The M98 in line N005 of the main program tells the control to jump to program number 1000. Program number 1000 is the bar-feed subprogram and remains permanently within the control's memory. During setup, the operator will modify one command of this program (the Z position of line N015) in order to describe the length of the next workpiece to be machined.

Spindle-speed considerations

Today's CNC turning centers allow very fast spindle speeds. Over 6000 RPM is available on many current models. However, in bar feeding, the weight, length, and straightness of the bar dramatically affect the maximum spindle speed that can be achieved without excessive vibration.

For example, it is likely that a 2-in-diameter, 12-ft-long bar cannot be rotated much faster than about 2500 RPM, even with a good-quality hydraulic-style bar feeder. Yet if constant surface speed is used, there may be many times when the spindle will attempt to run faster than the maximum safe speed. Facing a workpiece to center is one time when the spindle will attempt to run up to its maximum RPM. If the machine is capable of rotating faster than 2500 RPM, there will be a vibration problem. For this reason, most turning-center control manufacturers allow the programmer a way to specify the maximum allowable RPM within the program. Most use a G50 to accomplish this. Once the G50 command is executed, the control will not allow the spindle to run faster than the G50 command specifies.

The G50 command includes an S word to specify the maximum RPM to be allowed for the program. The command

N005 G50 S2500

for example, tells the control not to let the spindle run any faster than 2500 RPM, even if it is commanded (whether it is commanded directly with an RPM spindle-speed designation or indirectly through the constant-surface-speed mode).

To determine the maximum allowed RPM, the operator performs a simple test. He or she will load the new bar and start the spindle at a rather slow RPM. Then, in small increments, they will command that the spindle run faster and faster. When the vibration becomes excessive, the maximum spindle speed has been reached. To allow a safety margin, most operators will reduce this speed by about 10 percent.

Spindle-speed limiting versus machining time.   Unfortunately, there will be many times when the maximum spindle speed without vibration is much slower than that recommended for optimum cutting conditions. In most cases, running the spindle too slow will have no adverse affect on workpiece quality, but cycle time will obviously suffer. For high production quantities, this can be a big problem. For example, if spindle speed must be compromised to only 50 percent of optimum, it will take twice as long as it should to perform the machining operation!

For low production quantities, most companies simply live with this spindle-speed compromise. But when production quantities grow, there is a way to minimize the lost time due to inefficient spindle speeds. It will take some time to develop this technique for each new job, but once developed, this technique can save countless hours of production time.

Say, for example, you are running a 2-in-diameter, 12-ft bar. Say that each workpiece to be machined is 1 in long (including cutoff tool width and facing stock). In this case, approximately 140 workpieces will be machined per bar.

When the bar is first placed into the spindle, the bar is, of course at its heaviest. Say you test for vibration and determine that the maximum spindle speed for a new bar is 2500 RPM. But as workpieces are machined, the weight of the bar will drop, and it is very likely that this maximum RPM can be increased without inducing vibration.

Let's say you decide to break the bar up into four segments (you can make as many segments as you wish). In our case, each segment is 35 pieces (140 divided by 4). For the first bar you run, simply stop the cycle after 35 pieces have been run.

Now test for maximum RPM again. It is likely that you can now run the spindle somewhat faster than 2500 RPM. Say, for example, you find it can now run at 3200 RPM without vibration. Document this new maximum RPM.

Next, run 35 more workpieces and stop again. Test for your new maximum RPM, and document the new maximum. For our example, let's say it comes out to 3800 RPM.

Finally, run 35 more workpieces, stop, and test for maximum RPM one last time. This will be the final maximum for the balance of your bar. We'll say it comes out to 4500 RPM.

With this new knowledge of how your bar can be run, let's look at a simple program that takes advantage of faster speeds as the bar length shortens.

```
O0002 (Main program)
N005 G50 S2500 (Limit maximum RPM for first segment)
N010 M98 P1000 L35 (Run the first 35 workpieces)
N015 G50 S3200 (Limit maximum RPM for second segment)
N020 M98 P1000 L35 (Run the second 35 workpieces)
N025 G50 S3800 (Limit maximum RPM for third segment)
N030 M98 P1000 L35 (Run the third 35 workpieces)
N035 G50 S4500 (Limit maximum RPM for fourth segment)
N040 M98 P1000 L35 (Run the fourth 35 workpieces)
N045 M30 (End of program)
O1000 (Subprogram)
N005...(Bar-feed and machine entire workpiece)
.
.
.
M99 (End of subprogram)
```

As you can see, the main program simply sets the maximum RPM and runs 35 workpieces. Then the maximum RPM is increased (based on the previous testing) and 35 more workpieces are run. This process is repeated for the entire length of the bar.

Admittedly, this technique does require some testing during setup. But for high production quantities, the time it takes to do the testing can be easily retrieved during the first few bars you run! Also, once you determine what your maximum RPM segments will be for a particular bar size, they will remain the same for that bar size even for the next bar. This means you should not have to test maximum RPM for every setup, only once for every new bar size.

There are those who will argue that the workpiece quality will change during the spindle RPM changes. Workpieces run at slower spindle speeds will have different witness marks than those run at higher speeds. We contend that those workpieces run at higher speeds will actually be of better quality, since a speed closer to the correct speed is being used. However, if there can be no inconsistency among workpiece finish for all of the workpieces, you may be forced to run all workpieces at the first (slow) spindle speed.

Keep in mind that this technique makes an excellent application for parametric programming. Once you have made your RPM tests, you can let the parametric program do the math necessary to determine how many segments you wish, as well as how many workpieces there are per segment, based on the length of the bar and workpiece.

Tooling considerations

As you can imagine, since the CNC turning center will be unmanned during the machining of all workpieces from the bar, the tooling (and cutting conditions) must be designed accordingly. If the machine utilizes an automatic bar loader, this problem is compounded since each tool may have to last for a very long time without operator intervention.

This situation makes an excellent application for a tool-life management system. Tool-life management systems for turning centers (discussed later in this chapter) allow the programmer to set up a group of identical tools for each machining operation. As one tool wears out, the next tool in the group becomes the active tool.

Part Catchers

When a workpiece is cut off during a bar-feed operation, it will fall from the machining area. If no special provisions are made, the workpiece (still rotating at a very fast rate) will fall to the bottom of the machine's bed, possibly into the chip conveyor. During this fall the workpiece will bounce around quite a bit due to the inertia of its rotation. It is *very* likely that the workpiece will be damaged during this fall.

Figure 3.22 Part catcher attached to a CNC turning center. (*Courtesy Mori Seike USA.*)

Some companies attempt to solve this problem by padding the entire lower internal surfaces of the machine. Though this may help keep the workpiece from becoming damaged, the operator will eventually be digging around in the chip bin in order to find the workpieces.

A part catcher eliminates this problem. Part catchers swing up to catch the workpiece just before cutoff. After cutoff, the workpiece falls gently (one hopes) into the part catcher. As the part catcher retracts from the work area, it drops the workpiece into a collecting bin. Figure 3.22 shows a picture of a CNC turning center with a part catcher.

The motions of the part catcher are programmable by M codes. One M code commands the part catcher to swing into position and another retracts the part catcher.

Note that part catchers vary dramatically in design from one turning-center manufacturer to the next. Also, keep in mind that their effectiveness also varies. Some will effectively catch well over 95 percent of workpieces cut off, while others catch a much smaller percentage.

Bar Pullers

If a CNC turning center is purchased exclusively for bar work, the best possible feeding device is a (hydraulic-style) bar feeder. A bar machine should also be equipped with a collet chuck for work holding and a part catcher to remove workpieces from the work area.

However, as with all facets of the machining environment, it is sometimes necessary to make compromises. It is not always possible to set up ideal machining conditions. For example, if a company has a variety of work to be machined on a turning center, it may not be feasible to purchase a single-purpose machine. The machine may have to do chucking work and shaft work, as well as bar work.

In this case, compromises will have to be made when it comes to using the machine for bar work. Probably a general-purpose three-jaw chuck will be equipped with the machine instead of a collet chuck. Standard hard jaws or soft jaws bored completely through the jaw will be used to grip the workpiece instead of a collet. The machine will probably not be equipped with a part catcher. Some form of makeshift workpiece collector box will have to be fabricated. And finally, it may not be possible to justify the purchase of an expensive bar feeder if the percentage of bar parts to be machined by the turning center is small.

A bar puller is an inexpensive alternative to purchasing a bar feeder. Instead of pushing the bar through the spindle as a bar feeder does, a bar puller is placed in a turret station. This device locks on to the bar. The chuck is then opened and the turret is moved in the plus Z direction, pulling the bar through the spindle. The chuck is then closed, the bar puller retracted, and the cycle continues.

Though bar pullers vary from one manufacturer to the next, most employ some form of spring-loaded clamping device to grip the workpiece during the bar pull. This clamping must be quite positive to ensure a precise bar pull distance, especially for heavy bars. One popular configuration resembles a pair of pliers. This form of bar puller has serrated teeth that are parallel with the machine's X axis and engages the workpiece during an X-minus movement. During this engagement, the serrated teeth actually bite into the outside diameter of the bar to some extent, providing a very positive grip. When disengaged, this bar puller must be retracted in the X plus direction. Figure 3.23 shows a drawing of this popular form of bar puller.

Bar puller limitations

Though bar pullers offer a very economical way of performing bar work, there are several limitations of which you must be aware. The first is the length of the bar that can be machined. Unless some form of bar support system is fabricated, the rough stock bar length will be limited to the length of the spindle. *Under no circumstances* can an operator let the bar extend unsupported from the back of the spindle

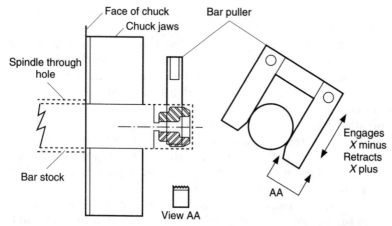

Figure 3.23 Bar puller for turning-center use. Note serrations in grippers that will bite into workpiece.

(the side of the spindle opposite the chuck). Doing so creates a *very dangerous* situation.

Here's why. Say, for example, a piece of 1-in bar stock extends 2 ft from the back of the spindle (unsupported). Say that during some point in the machining cycle, the spindle will run at a high spindle speed, say 1500 RPM. The unsupported end of the bar will be under a great deal of stress at this point and will instantaneously bend to a 90° angle due to centrifugal force caused by the high-speed rotation. Anything in the way of the bar will be damaged, including machine covers, spindle motor, and possibly the operator. *Never* allow an unsupported bar to extend past the end of the spindle!

This means the effective length of bar stock that can be safely machined will be equal to the spindle length. For most turning centers, the spindle length is under about 4 ft. This means that bars up to only about 4 ft can be machined. If bars are supplied in 12-ft lengths, a sawing operation must be performed prior to the CNC turning-center operation, meaning more workpiece handling.

This limitation dramatically reduces the feasibility of performing bar operations. One of the most basic reasons for bar-style machining in the first place is reduced operator intervention. If only 3- to 4-ft bar lengths are allowed, and if the workpieces to be machined are relatively long, only a few workpieces may be machined per bar.

Note that even when the bar is totally enclosed in the spindle, it will still have the tendency to rattle around at high speeds. This means the operator must cautiously check for maximum possible spindle speed before machining can be done (as was discussed in the presentation on bar feeders). It is likely that a severe spindle-speed limitation will have to be imposed within the program, which, in turn, increases cycle time.

Another limitation of bar pullers is that they take more time per bar pull than bar feeders. With a bar feeder, the cutoff tool can be used as the stock stop. Since the cutoff tool is still in position and ready as soon as the workpiece is cut off, the bar feed can occur immediately.

On the other hand, when you are using a bar puller, the cutoff tool must be sent to a tool-change position. The turret must then be indexed to the bar puller station. Finally, the bar puller must be brought back to the workpiece. All of this must happen *before* the bar pull can begin, and of course, takes precious cycle time.

This brings up the next limitation. When a bar feeder is used, all tools in the turret can be cutting tools. In bar pulling, a turret station *must* be used for the purpose of holding the bar puller.

Another serious limitation of bar pullers is that they leave witness marks on the bar. Since the bar puller teeth are serrated, they gouge into the bar to some extent. As long as the outside diameter of the bar is to be machined during the machining operation (as the example in Fig. 3.23 shows), this presents no major problems. On the other hand, if the outside diameter is not to be machined, the witness marks left by the bar puller may make the workpiece unusable.

Since the bar puller depends on spring-loaded serrations to grab the workpiece and provide positive gripping, hard and smooth materials (like some forms of ground stock) make it difficult to bar-pull. Instead of forming a positive grip on these materials, the bar puller tends to slide along the surface of the bar during the pulling motion, especially for heavier bars.

When a bar puller is used, the operator must load new bars precisely, since no solid stock stop is used. If the bar is slightly out of position with respect to the face of the jaws, there will be too much or too little stock to be removed from the face of the first workpiece.

The final limitation we will mention has to do with the fact that there will be *no* end-of-bar signal. Since there is no (feasible) way to interface the bar puller with the control (as a bar feeder is interfaced), the programmer and operator *must* be *very careful* with regard to how many workpieces will be machined per bar. If a mistake is made in this regard, the final workpiece on the bar may not have enough chucking surface, and the results could be disastrous.

Bar puller programming

Most of the same basic points about bar-feeder programming still apply to bar pulling. Many of the same commands are required, like chuck open and close, inches per minute versus inches per revolution mode, etc., and the considerations discussed about when to perform the bar pull (at the beginning or at the end of the program) still apply. However, there are two important differences.

One difference has to do with how the bar puller engages and drags the bar. Most bar puller manufacturers provide suggestions for how the puller should be programmed.

The second difference has to do with how many workpieces can be machined per bar. Since there is no end-of-bar signal as there would be with a bar feeder, the programmer *must* synchronize the program with the number of workpieces that can be safely machined per bar. This, of course, means bar lengths must be kept quite consistent, allowing the same number of workpieces per bar to be machined.

A relatively easy way to handle this number of workpieces programming problem is to use subprogramming techniques. A main program can simply call the subprogram command to be activated a specified number of times. This main program then ends in the normal manner, stopping the cycle. The subprogram completely machines the workpiece, and can include the bar pulling routine. Of course, if bar pulling is done often, the bar pulling commands can also be stored as a subprogram (just as was discussed for bar-feeding commands).

Let's take a look at an example. For our example, we'll use the style of bar puller shown in Fig. 3.23. We'll say that 15 workpieces can be machined per bar and that the bar must be pulled precisely 0.905 in per pull. Program zero is set at the right end of the finished workpiece. To keep the bar pull simple, the incremental mode will be used once the bar puller engages the workpiece.

Main program:

O00001 (Program number)
N005 M98 P1000 L15 (Execute subprogram number 1000 fifteen times)
N010 M30 (End of program, cycle stops)

Subprogram:

O1000 (Subprogram number)
N005 T0101 (Index the turret to the bar puller)
N010 G00 X2.0 Z-0.625 M05 (Rapid to a clearance position above the workpiece, ensure that the spindle is off)
N015 G98 G01 X-1.5 F25. (Select IPM mode, fast-feed the bar puller to engage the workpiece)
N020 M15 (Open jaws)
N025 W0.905 F30. (Incrementally, pull the bar plus in Z by 0.905 in)

N030 M14 (Close jaws)
N035 X2. (Retract the bar puller)
N040 G99 G00 X6. Z5. (Reselect IPR mode, rapid to tool-change position)
...(Machine workpiece and cut off)
...M99

Automatic Tool-Changing Systems

Most turning centers are equipped with a single turret that can hold up to 10 or 12 tools. For most applications, this is an adequate number of tools. However, there are times when more tools may be needed—for example, with flexible manufacturing systems and automation systems requiring the machine to be tooled up for several different workpieces. Though not every tool may be needed for every job, the turning center must have the tools available for automatic retrieval.

To a great extent, automatic tool changers for turning centers resemble those for machining centers. A tool carousel, or magazine, that can hold even more than 20 or 30 tools works in conjunction with the machine's turret. The tools in the magazine can be quickly and automatically exchanged with tools in the turret.

Since tool changing must be totally automatic, an automatic tool-changing system requires some form of automatically activated tool-clamping device within the turret. Tools are clamped into the turret in much the same manner as rotating tools are clamped into the spindle of a machining center. In fact, one popular style of turning-center tool holder shank very much resembles the tapered shank style used with machining-center tooling.

Programming considerations

Though the programming commands for turning-center automatic tool changers will vary from one manufacturer to the next, most use techniques that are quite similar to those used for a machining center's automatic tool changer. A series of two-digit T words are used to specify magazine positions. An M06 actually makes the command to change tools, exchanging the tool in the magazine's ready position with the tool currently in the turret's ready position. Since a two-digit T word is used for magazine rotation, the control can distinguish it from the (four-digit) T word used for turret indexing.

Note that most machines do *not* load the tool into the working turret position. Most load tools into a position that is more convenient for tool loading and free of obstructions. For example, when turret station 1 is in the machining position, turret station 3 may be the turret station that will accept a tool from the magazine. While this can be a little confusing to the programmer, the machine-tool builder may have no other way to handle this tool-loading problem.

Like machining-center automatic tool changers, most turning centers require that the machine be positioned to a precise tool-change position before a tool change is possible. Though this position may vary from one builder to the next, most use the machine's reference position as the automatic tool-changer position.

Here's an example. Say that the tool in turret station 5 is to be replaced with the tool in magazine position 18. Say the turret station number that is the tool-changing position is 3 plus the turret station number in the cutting position. Here is a series of commands that would complete the tool change.

Program:

O0001 (Program number)
N005 G28 U0 W0 (Send the machine to its reference point)
N010 T18 (Rotate the magazine so that tool number 18 is in the ready position)
N015 T0200 (Rotate the turret to station number 2; this puts turret station 5 in the ready position)
N020 M06 (Change tools)
...
N445 M30 (End of program)

Keep in mind that the basic intention of an automatic tool changer for a turning center is slightly different than that for a machining center. With a machining center, only one tool is available at a time to the machine for cutting purposes (the tool in the spindle). By comparison, a turning center will have as many as 12 tools, all available for machining. Since most workpieces can be totally machined with less than 12 tools, the automatic tool changer will only be required as the machine goes from machining a workpiece in one setup to machining another in a different setup. Since this kind of tooling flexibility is almost exclusively required in flexible manufacturing systems, automatic tool-changing systems are seldom found on stand-alone turning centers.

Due to the complexity of this kind of system, and due to the sheer number of tools available to the CNC turning center when an automatic tool-changing system is equipped, some form of tool-life management system is normally required to manage the various tools in the system. Tool-life management systems as they are applied to turning applications are discussed later in this chapter.

Live Tooling

The primary use for any turning center is to machine workpieces as an engine lathe does. A stationary cutting tool is brought into contact with a rotating workpiece. Operations such as rough and finish turning, rough and finish boring, grooving, knurling, and threading are performed in this fashion. Until relatively recently, a turning center was limited to functioning only in this manner.

Many workpieces to be machined on a turning center require secondary operations after the turning operation. A flange, for example, may require that a series of holes around a bolt-hole pattern be machined after the turning is done. A shaft may require a keyway to be milled or a cross hole to be drilled. In some cases, a rather complex contour must be milled around the periphery of a specific diameter.

Traditionally, these secondary operations are performed on separate machine tools like manual milling machines and drill presses *after* the turning center operation. Each workpiece requiring secondary operations is brought to another machine for additional machining. As you can imagine, secondary operations mean more setups and more workpiece handling. This equates to higher production costs. For this reason, more and more companies are attempting to minimize (or eliminate) secondary operations by performing the necessary secondary operations on the turning center itself. When this is possible, not only are production costs reduced, but workpiece quality also improves because fewer setups are required.

In order for the turning center to perform secondary operations, it must have the ability to perform milling, drilling, tapping, and other operations *not* commonly considered as turning-center operations. There are those in this industry who oppose the use of a turning center for operations other than turning operations. These people say that you should use the machine tool for the purpose it was designed for. They say a turning center is not generally designed to perform other forms of machining operations and will not be rigid enough for heavy machining.

Keep in mind, however, that, for the most part, secondary operations do not require a great deal of power. If only light-duty secondary machining operations are to be performed, a well-designed turning center can accomplish them with ease. However, a turning center must be equipped with several special and additional features in order for secondary operations to be possible.

First, the turning center must be able to rotate a tool held in the turret (hence the name *live tooling*). A special spindle drive motor and transmission mounted within the turret provide the necessary tool rotation. The speed of the rotating tool must, of course, be a programmable function.

Second, the tool holders that contain rotating tools must have the ability to point the tool along the X axis (as would be required for drilling cross holes) as well as along the Z axis (as would be needed for drilling flange holes). Figures 3.24 and 3.25 show examples of how rotating tools are held in the turret.

These tool holders must hold the tool rigidly without causing excessive interference problems. Note that turning-center manufacturers that provide the live tooling option vary with regard to how well the interference problem is handled. Some allow any tool station to be used as a live tooling station and present a minimum of interference problems. Others limit the number of tool stations that can hold rotating tools, and their tool holders themselves create an interference nightmare. With the worst-designed live tooling turning centers, each tool station adjacent to the live tool station must be left empty to provide the necessary clearance for the rotating tool and holder. This, of course, dramatically reduces the number of effective tool stations available for machining.

Third, the programmer must have precise control of the main spindle's rotation. During any normal turning operation, the machine's spindle is simply rotated at a specified speed. When the main spindle is stopped, it is in a neutral mode, still rather free to rotate. There

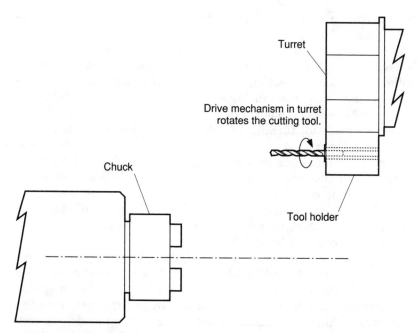

Figure 3.24 Drawing illustrates a live tool being held parallel to the Z axis. This kind of tool is used for end work.

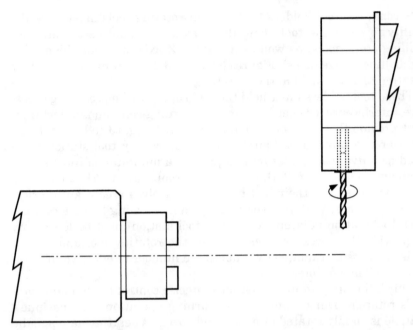

Figure 3.25 Drawing illustrates a live tool being held parallel to the X axis. This kind of tool is used for diameter work.

must be some provision for locking the main spindle in an accurate and programmable manner.

There are two common ways turning-center manufacturers handle this problem. One way is to utilize an indexing device within the spindle. The other way is to incorporate a full rotary axis within the spindle. Either way, the machine-tool builder must provide a way (usually by M codes) to specify which spindle mode is desired—the free turning mode for turning operations or the indexer/rotary mode for live tooling operations.

The most common form of indexing device used for this purpose is a 1° indexer. If the rotary device within the spindle is a true rotary axis, this axis is called the C axis. These rotary devices are programmed in much the same way rotary devices for machining centers are programmed. More on programming rotary devices for turning-center applications a little later.

Fourth, whenever live tools are used, spindle speed (for the live tool) must be programmed in RPM mode. Since the rotating tool's diameter is constant, the feature constant surface speed is not applicable. Along the same lines, feed rate *cannot* be programmed in inches per revolution. Since the spindle is not in the turning mode, the control will consider the spindle off. If a movement is given with an inches per revolution feed rate, the machine will not move.

How feed rate *is* programmed depends on what kind of rotary device is being used within the spindle. If an indexer is used, *all* feed rates will be in inches per minute in the inch mode or millimeters per minute in the metric mode. If a true rotary axis is used, and if only X and/or Z motion is required, feed rate can also be specified in inches per minute or millimeters per minute. However, any feed-rate motion including a rotary axis departure *must* be programmed in degrees per minute (DPM).

Fifth, if a great deal of side milling is to be done, it will be helpful if the CNC control is equipped with the feature cutter radius compensation (as on machining centers). This will make it very easy to compensate for the milling cutter's diameter. This feature is especially helpful when a contour must be milled around the periphery of the workpiece.

Sixth, most turning centers with live tools are equipped with a full set of hole-machining canned cycles for center drilling, drilling, tapping, reaming, and counterboring. These cycles closely resemble those used for machining-center hole-machining operations.

As you can see, a turning center equipped with the live tooling option has several features a normal turning center would not have. These features, of course, dramatically increase the cost of the machine. But if secondary operations are often performed, the features can be easily justified when you consider the savings in production time and improvement in workpiece quality.

Programming the rotary device

As stated, the rotary device that is internal to the spindle of a turning center with live tooling is programmed in much the same way any machining-center rotary device is programmed. If you have read the presentations made in the previous chapter, some of the information presented in this section will be a little redundant.

Specifying the main spindle mode. *All* turning centers that are equipped with live tooling must allow some way to let the programmer specify which spindle mode is desired—the normal turning mode or the rotary-device mode. Two M codes are usually used for this purpose.

One M code selects the normal turning mode. This mode is usually initialized; that is, at power up, the machine automatically selects the turning mode. In this mode, the spindle functions like that of any normal turning center.

Another M code selects the rotary device mode. When this M code is executed, some form of mechanical device within the spindle housing will engage the rotary device. Keep in mind that this M code will usually perform some other functions as well. When the rotary device mode is selected, the spindle-related commands will be transferred to the live tool spindle drive motor.

When the rotary device mode is selected, for example, any spindle speed word (specified by an S word) will be taken as the speed used for the live tooling spindle drive motor. M03 will turn the live tooling spindle drive motor on in the forward direction. M04 turns it on in the reverse direction, and M05 stops the live tooling spindle drive motor. Note that *only* the RPM mode is allowed with live tooling. Constant-surface-speed mode is not allowed, and most controls will generate an alarm if it is selected for use with live tooling.

Indexers. An indexer is used to quickly rotate the workpiece to an attitude that allows a machining operation to be performed. *No* machining can occur during the index. The drawing shown in Fig. 3.26 shows a good application for an indexer.

Notice the series of holes to be machined in the face of this workpiece. The indexer within the spindle can be easily commanded to rotate the workpiece to the first angular hole location. One hole can then be machined. Then the indexer can be commanded to rotate 45° to the next hole location. Another hole can be machined. This process can be repeated for all eight holes.

Keep in mind that with an indexer, machining can only occur *after* a rotation. And since the main spindle is stopped during any live tool-

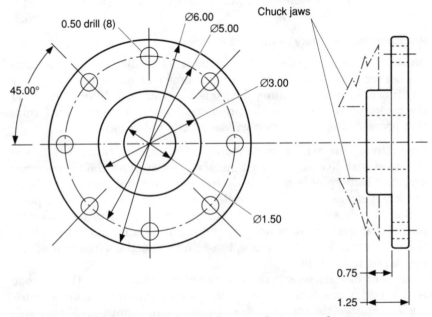

Figure 3.26 Workpiece made by indexer-style rotary-device example program.

ing operation, *all* feed rates *must* be programmed in the inches per minute (or millimeters per minute) mode.

Indexers equipped with most CNC turning centers are 1° indexers. One-degree indexers use a special programming word with which to command the angular departure distance. The letter address C is often used for this purpose. With this kind of indexer, to index 45°, this command is given:

N040 C45

Most turning centers with indexers allow the programmer to specify the direction of rotation within the indexing command. Most use two M codes to control the direction of rotation. One M code specifies clockwise rotation while another specifies counterclockwise rotation.

Though not as common, the rotary device on some turning centers is a 5° indexer. These indexers usually have a series of M codes to specify the possible indexing angular amounts. M71, for example, may command a 5° index. M72 may command a 15° index; M73, a 30° index; and so on.

Example program for an indexer. This example program will perform the drilling operation for the workpiece shown in Fig. 3.26. For the purpose of this example, we'll say that M81 is the M code that selects the normal spindle rotation mode and M82 is the M code that selects the indexer mode. We'll say the machine has a 1° indexer and that it is commanded by a C word.

Program:

O0001 (Program number)
N005 M81 (Select turning mode)
...(Completely turn workpiece)
...N205 T0101 (Rotate turret to 0.500-in drill)
N210 M82 (Select indexer mode)
N215 G97 S650 M03 (Turn the live tool spindle motor on at 650 RPM in the forward direction)
N220 G00 X5. Z.1 (Rapid to first hole location)
N225 G81 R.1 Z-.65 F5.0 (Drill first hole)
N230 C45 (Index 45° and machine another hole)
N235 C45 (Index another 45° and machine another hole)
N240 C45
N245 C45
N250 C45
N255 C45
N260 C45 (Machine last hole)
N265 G80 (Cancel drilling cycle)
N270 G00 X10. Z5. (Return to tool change position)
N275 M81 (Reselect normal turning mode)
N280 M30 (End of program)

Notice that this program used the canned cycle for drilling to make programming easier. Once instated, this canned cycle (G81) will continue to machine a hole in each command until it is canceled by the G80.

Rotary axes. When a rotary axis is equipped within the spindle of a CNC turning center, it is called the *C axis*. As with any rotary axis, the programmer will have total control of rotary-axis departures. A rotary axis is very similar to any (linear) axis with regard to programming techniques. Some of these similarities include the designation of program zero, rapid and straight-line cutting commands, and the use of incremental and absolute motion modes.

Absolute versus incremental. A rotary axis can easily be used as an indexer. The index is simply commanded by a G00 command. The easiest way to command indexing on a rotary axis is with the incremental mode. For example, the command

    N050 G91 G00 C45.

will simply rotate the workpiece 45° in the clockwise (plus) direction. The command

    N055 G91 G00 C-45.

will rotate the workpiece 45° in the counterclockwise (minus) direction.

Note that our two example commands include the G91 word to specify the incremental mode. However, some turning centers do not utilize G90 and G91 to specify absolute and incremental modes. Instead, the letter address related to the axis departure designates whether the motion is to be made in the absolute or incremental mode. For example, many machining centers use the X and Z letter address to designate that the motion is to be made in the absolute mode. The command

    N035 G00 X3. Z.1

tells the control to rapid the tool to an absolute position of 3.0 in X and 0.1 in Z.

The letter addresses that designate incremental mode motions for this type of machine are U for incremental X-axis motions and W for incremental Z-axis motions. The command

    N040 G00 U1. W3.

tells the control to move the tool from its current position by a diameter increase of 1 in (0.500 in actual motion) in the X axis and 3 in in the Z axis.

When incremental movements are specified in this manner, the letter address commonly used to specify incremental C-axis departures is the letter address H. The command

N035 G00 H-45.

tells the control to incrementally rotate the workpiece clockwise (plus) by 45°. H-45. specifies a counterclockwise (minus) incremental rotation.

Rapid versus straight-line motion. As you know, it is relatively simple to use the rotary device as an indexer by causing rotation with a rapid command (G00). However, if the turning center has only three axes (X, Z, and C), performing any machining operation *while* the C axis rotates can be quite tricky.

Say, for example, the flats shown in Fig. 3.27 must be machined by an end mill held parallel with the Z axis. Unless the turning center is equipped with a Y axis (and most are not), this kind of milling requires a combination of X-axis linear motion with C-axis rotary motion. Figure 3.28 demonstrates this combination of X and C motion on a hex-shaped workpiece.

A beginner, exposed to this kind of problem for the first time, may jump to a wrong conclusion. The beginner may think that each flat could be machined in two commands, one to command the end mill to move from one end of the flat to the center of the flat, and another to

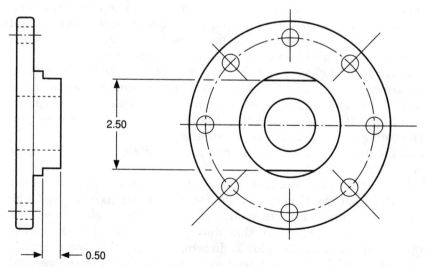

Figure 3.27 Some workpieces require milling operations that must be performed around their diameters.

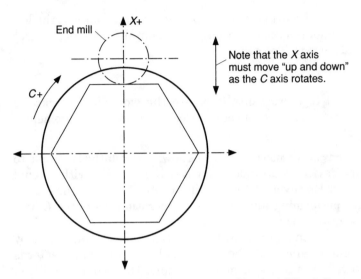

Figure 3.28 This kind of milling requires a combination of *X*-axis linear movement along with *C*-axis rotation.

command the end mill to go to the opposite side of the flat. Unfortunately, it is not that simple. For every degree (or portion of a degree) of rotation, a different amount of *X*-axis motion is required.

The first CNC turning centers to be equipped with a *C* axis and live tooling were extremely difficult to program in this regard. In all cases, parametric programming or some form of computer-aided manufacturing system was required in order to generate the commands needed to machine flats on a workpiece. And if any form of circular command was involved in the series of milling movements, these older machines were next to impossible to program.

Understanding polar coordinate interpolation. In order to make programming simpler for this kind of machining operation, most turning-center control manufacturers offer a feature called *polar coordinate interpolation.* When this feature is used, it makes programming any flat or radius as simple as giving one command. To mill the hex-shaped workpiece shown in Fig. 3.28, for example, takes only six commands!

In essence, polar coordinate interpolation allows the programmer to flatten out the *C* axis and view it as a linear axis. Figure 3.29 shows how the coordinate system is viewed when polar coordinate interpolation is applied. In this view, you are looking at the end of the workpiece from the plus *Z* direction. Notice that the *C* axis is viewed as the horizontal linear axis and the *X* axis is the vertical axis. Notice also that any *C* coordinate to the right of the program

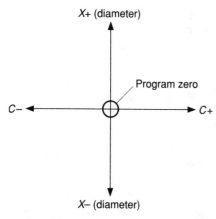

X+ (diameter)

Program zero

C–   C+

X– (diameter)

Figure 3.29 Drawing illustrates the coordinate system used with polar coordinate interpolation. Note how the C axis can be viewed as a linear axis when polar coordinate interpolation is used.

zero point is plus and any C axis coordinate to the left of program zero is minus. Any X coordinate above program zero is plus and any X coordinate below program zero is minus. Keep in mind that *every* X-axis coordinate is specified in diameter. This sometimes means (radius) print dimensions must be doubled.

Also related to the X-axis coordinate designation, there will often be times when the coordinate specified must be quite a bit minus. This tends to confuse the beginning C-axis programmer a great deal, since, in the normal turning mode, X-axis designations are seldom minus. When polar coordinate interpolation is used, however, it completely changes the way the programmer must view the X and C axes.

When polar coordinate interpolation is used, each surface of the workpiece can be machined in one command (instead of hundreds of commands). This also includes surfaces that must be machined by circular motions. Some versions of polar coordinate interpolation even allow cutter radius compensation, just as it is used on a machining center. For these controls, calculating coordinates in X and C is very simple, since the tool's centerline coordinates need not be considered. In fact, for programming purposes, the programmer need not even know the precise diameter of the milling cutter to be used.

Other versions of polar coordinate interpolation do not allow cutter radius compensation techniques. These controls force the programmer to calculate coordinates based on the milling cutter's centerline coordinates. This, of course, means the programmer *must* know the diameter of the milling cutter and use its size during coordinate calculations.

Figure 3.30 shows a rather complicated workpiece contour that includes circular motions. A coordinate sheet is also given, showing how coordinates are calculated. For this example, note that cutter radius compensation is not being used, meaning the milling cutter's

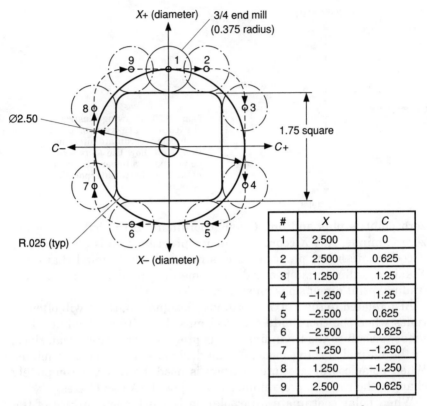

Figure 3.30 Example workpiece for polar coordinate interpolation example program. Coordinate sheet illustrates how coordinates are calculated when polar coordinate interpolation is used.

#	X	C
1	2.500	0
2	2.500	0.625
3	1.250	1.25
4	−1.250	1.25
5	−2.500	0.625
6	−2.500	−0.625
7	−1.250	−1.250
8	1.250	−1.250
9	2.500	−0.625

centerline coordinates are given for a 0.750-in-diameter (0.375-in-radius) end mill.

You may be questioning some of the X coordinates related to this coordinate sheet. Keep in mind that *all* X coordinates *must* be given as diameter values. Since these coordinates represent the cutter's centerline coordinates, all coordinates must take into consideration the 0.375-in radius of the cutter. To calculate the X value for point 2, for example, requires that the 1.75-in distance across the flats be added to the tool diameter (0.750 in) in order to come up with the 2.500-in X value.

Notice that some X coordinates are quite negative. Though the X axis of the machine will not truly move this far past center, polar coordinate interpolation requires the programmer to enter X coordinates in this manner.

With these points understood, let's look at an example program for the workpiece shown in Fig. 3.30. Note that this particular control uses a G112 to instate the polar coordinate interpolation mode and a G113 to cancel it. Also note that the machine for which this program is written uses an M81 to instate the normal turning mode and an M82 to instate the rotary axis mode.

O0001 (Program number)
N005 T0101 (Index the turret to the ¾-in end mill)
N010 M82 (Instate the rotary device spindle mode)
N015 G112 (Instate polar coordinate interpolation mode)
N020 G00 X3.65 C0 (Rapid to clearance position)
N025 G98 G01 X2.5 F10. (Select IPM mode, feed to point 1)
N030 G01 C.625 (Feed to point 2)
N035 G02 X1.25 C1.25 R.625 (Form radius to point 3)
N040 G01 X-1.25 (Feed to point 4)
N045 G02 X-2.5 C.625 R.625 (Form radius to point 5)
N050 G01 C-.625 (Feed to point 6)
N055 G02 X-1.25 C-1.25 R.625 (Form radius to point 7)
N060 G01 X1.25 (Feed to point 8)
N056 G02 X2.5 C-.625 R.625 (Form radius to point 9)
N070 G01 C0 (Feed back to point 1)
N075 G00 X3.65 (Rapid away in X)
N080 G113 M81 (Cancel polar coordinate interpolation mode, reinstate normal turning mode)
N085 G00 X10. Z5. (Return to tool-change position)
N090 M30 (End of program)

As you can see, only nine commands were required to actually mill the outside square shape. With polar coordinate interpolation, notice how easy it is to form each surface. Even circular motions that form the radii between straight surfaces are very easy to program! As compared to the old way (using parametric programming techniques or a CAM system), this method offers a dramatic savings of time, effort, and CNC control memory requirements.

Components of Automation Systems

Automation systems are designed to minimize or eliminate the need for operator intervention during the running of a quantity of workpieces. Some automation systems even attempt to minimize the need for operator intervention during the changeover from one setup to the next.

As discussed earlier in this chapter, one form of automation system for turning centers incorporates a bar feeder. When it comes to complexity, bar-feeding automation systems are among the simplest to implement. This is attributable to the simplicity of the workpiece in its raw state. Bar stock of a limited number of shapes and sizes is machined.

Almost any other form of automation system will be *much* more complicated. Some form of loading device must be specially designed and fabricated. This loader must be designed with *all* workpieces to be machined by the turning center in mind. As you can imagine, this can be a very complicated task.

In this section, we cannot hope to present the criteria for how automation systems are developed. Indeed, the criteria change from one company to the next. What we can and will do is present those accessory devices that are almost always required with every form of automation system.

Tool-life management systems

The basic reason for using tool-life management systems on turning centers is the same as it is for machining centers: to minimize (or in some cases, eliminate) machine downtime while tools are being loaded. Instead of loading tools while the machine sits idle, tools are loaded during the machining cycle. A secondary benefit of tool-life management systems is prolonged duration of machining cycles during unmanned operation. The machine can be left unattended for long periods of time if duplicate tooling is used.

To truly reap the primary benefits of a tool-life management system, the turning center must also have an automatic tool-changing device. Since most turning centers have only about 12 turret stations, the turret by itself does not allow adequate storage for multiple tools. Automatic tool-changing systems allow many tools to be loaded into a tool magazine. They were discussed earlier in this chapter.

Keep in mind that without an automatic tool-changing system, tools must be loaded directly into the turret of the turning center. This means the machine must sit idle while tools are loaded. This defeats the purpose for using the tool-life management system in the first place, to minimize or eliminate the downtime caused during tool loading.

Note that tool-life management systems as they apply to machining centers are discussed at length in Chap. 2. How tool-life monitoring systems are programmed and utilized for turning centers is almost identical to how they are utilized for machining centers, so only additional considerations for turning-center applications will be discussed in this section.

As you know, tool-life management systems allow the user to double or triple the life of those tools that are most prone to failure. Instead of commanding that a single tool machine workpieces, a tool *group* is commanded to machine workpieces. This tool group will commonly contain several tools, with only one of the tools in the group being active at any one time.

Within each tool group, the operator will specify some form of tool-life duration designation. Usually this life duration is specified in time or in number of workpieces that can be machined. As an active tool wears out, the tool-life management system automatically indexes to the next tool in the group.

Like tool-life management systems for machining centers, each tool within a tool group has a separate tool offset to control the surfaces machined by the tool. But the tool offsets for turning centers work in a dramatically different way than those for machining centers.

Machining-center tool offsets simply specify the tool's length and diameter (or radius) value. These values can be easily measured off line. During the time of tool loading, the operator can easily enter the related tool offset values. Also, once a tool offset value is entered for a tool's length or radius, it will not change. The offset values will remain correct for the life of the tool.

Conversely, turning-center tool offsets are more dynamic. As you know, turning-center tool offsets control the size of the workpiece to be machined. If you are using conventional tool-setting methods, it can be next to impossible to truly determine what the value of a tool's offset should be before the tool machines a workpiece. This is specially true if close tolerances are to be held. Also, once a tool's offset value is determined, it does *not* remain the same for the life of the tool. As a turning tool wears, the value of the offset must be changed in order to keep the workpiece in size.

For these reasons, a tool-life management system by itself provides only part of what is required to effectively manage tools. It provides the tool grouping and offset grouping capabilities, but some other function must be available for checking workpiece size and updating the appropriate offset values. Tool point touch-off devices, in-process gauging probes, and postprocess measuring systems are examples of accessory devices that can be used for this purpose.

Probing devices

As with machining-center applications, probing systems for turning centers can be broken down into two distinctly different categories: tool touch-off probes and in-process gauging probes. Tool touch-off probes are used to help during setup, and in-process gauging probes are used to measure the workpiece during the machining cycle.

Tool touch-off probes.   As with automatic tool length measuring probes for machining centers, tool touch-off probes for turning centers help with locating the position of the tool tip. But instead of simply finding a tool length, this form of touch probe is capable of locating the

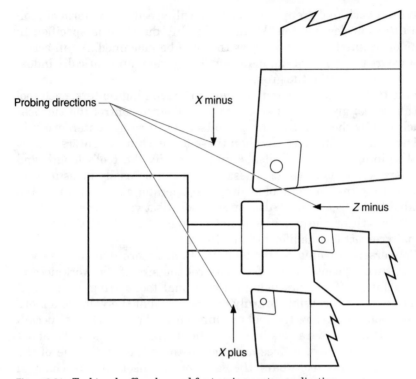

Figure 3.31   Tool touch-off probe used for turning-center application.

tool point in two directions, X and Z. Figure 3.31 shows a common touch probe used for turning-center application. Notice that it has three directions of probing, X minus, Z minus, and X plus.

Like touch probes used for machining-center application, a sensor within the probe sends a signal to the control the instant the probe is deflected. When using a probe for the purpose of tool point touch-off, the tool is driven into the probe along one axis. As soon as the tool contacts the probe, the probe sends the signal to halt axis motion. As with any probing system, there will always be a certain amount of overshoot. The machine cannot instantaneously halt the axis motion. This overshoot amount is determined during the calibration routine for the probe and must be taken into consideration every time a tool is touched. A great deal more about probe calibration is discussed in the probe presentation in Chap. 2.

Once the probe has stopped, the parametric program driving the probe can attain the current absolute position along the probing axis. This value can be used to calculate the value that must be placed in the tool's offset. In some cases, the offset is automatically inserted by the probing parametric program.

During the tool touch-off, the probe must, of course, be within the machine's travels in both axes. In many cases, the probe would interfere with the subsequent machining operations if left in this position. For this reason, most turning-center manufacturers that incorporate tool touch-off probing devices allow a way of swinging the tool in and out of the work area. While some manufacturers make the operator manually swing the probe into position, most make this function programmable (with two M codes). Note that the swing-down motion *must* be very accurate and repeatable. The probe *must* be positioned in precisely the same location each time it is brought into position if accurate tool touch-offs are to be done.

Some tool touch-off probe systems for turning centers require a great deal of manual intervention. With these probes, the operator must first manually swing the touch probe into position. Then the tool must be manually brought into close proximity to the probe in one of the axes. Next the operator will select the proper feed rate for touching the probe and manually jog the tool into the probe. Finally, once the probing operation is completed, the operator must manually input the position of the tool into an offset. This whole procedure must be repeated for the other axis of probing. And, of course this must be done for every tool in the setup. While this manual form of touch probe helps a great deal with setup, quite a bit of manual intervention is involved.

Other tool touch-off probe systems are totally automatic. As with automatic tool-length measuring systems used with machining centers, a standard program is used to specify which tool stations are to be probed. Also as with automatic tool-length measuring systems, the operator is required to specify a shift amount that tells the control the approximate location for the tool tip. Within the probing command, the operator will also tell the control what kind of tool is being probed (an outside diameter turning tool, a face profiling tool, or a boring bar). With this information, the probing program can automatically complete the probing operation for each tool and set offsets accordingly.

One limitation of tool touch-off probes on turning centers has to do with tool pressure. When the tool touches the probe, it is in a static condition, meaning no machining is taking place. As a tool actually machines the first workpiece, it will be under the influence of a certain amount of tool pressure. This tool pressure will sometimes dramatically affect the size of the surface machined by the tool.

Also, as it continues to cut workpieces, the tool will begin to wear, causing workpiece size variations. For these reasons, while the tool touch-off probe gives an excellent starting point for each tool, there must be some way of maintaining consistency from one workpiece to another. For relatively manual systems, the operator will measure each workpiece and adjust offsets accordingly. For automatic systems, some form of automatic gauging device (in-process or postprocess gauging device) must be used to keep offsets set properly.

In-process gauging probes.   An in-process gauging probe for turning-center application is mounted in a tool station on the turret of the turning center. When gauging is to be done, the turret is indexed to the probe station and the probe is commanded to move up to and contact the surface to be measured.

Virtually all of the points made during the discussion in Chap. 2 of in-process gauging probes will apply to turning-center in-process gauging probe applications. For this reason, we will not repeat the in-process gauging probe presentations in this chapter.

There is one dramatic limitation of turning-center in-process gauging probes that makes them almost infeasible to use. The machining environment of a CNC turning center is *extremely* messy. During machining, chips and coolant fly in all directions. If the in-process gauging probe is mounted in a tool station on the turret, it will be exposed to a great deal of grime. In extreme cases, the force of the flying hot chips may actually damage the sensitivity of the probe.

At the very best, the probe will eventually become quite dirty and incapable of taking accurate measurements. For this reason, most turning-center manufacturers are hesitant to equip their turning centers with in-process gauging devices unless an automatic tool-changer system is also purchased. With an automatic tool-changer system, the probe can be stored in the tool magazine, away from the messy machining environment.

Postprocess gauging devices

With any form of automation system, and especially for CNC turning centers, there *must* be a way to confirm that the workpieces being machined by the machine tool are making the workpiece to size. As you have seen, in-process gauging probes are slow, and work with only limited success for turning-center application.

By comparison, postprocess gauging devices work almost instantaneously and with a high degree of accuracy. Postprocess gauging devices most resemble an operator who measures every workpiece being machined. Like an operator, if the postprocess gauging device determines that a dimension is falling out of tolerance, it can easily and automatically update the corresponding tool offset. Figure 3.32 shows a postprocess gauging fixture designed to take two measurements.

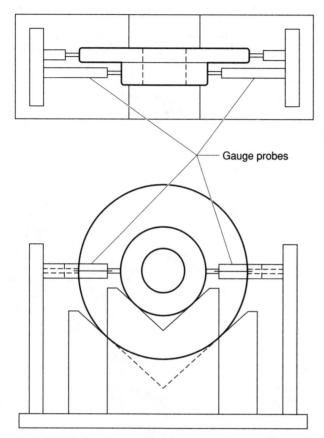

Figure 3.32 Postprocess gauging fixture used to measure critical dimensions on the workpiece.

Either by manual or automatic means, the machined workpiece must be placed in the gauging fixture. The postprocess gauging fixture must be designed to measure the dimensions of highest concern. In our relatively simple example, two outside diameters are to be measured. The more dimensions to be measured, the more complicated the postprocess gauging fixture must be. The gauging probes on the gauging fixture are connected to the postprocess gauging device's control system. The control system detects the positions of each probe to come up with the two dimensions being measured. Feedback to the CNC control in the form of tool offsets can be given if the dimensions are not precisely in the middle of their tolerance band.

As you can imagine, the postprocess gauging fixture can be quite complicated to design and fabricate, especially if many workpiece dimensions require measuring.

Index

ABOUT THE AUTHOR

Mike Lynch is president of CNC Concepts, Inc., a supplier of training materials and computer software to users of CNC equipment. He also conducts a seminar on CNC programming for the Society of Manufacturing Engineers (SME). Mr. Lynch previously worked as NC Operations Manager for K.G.K. International Corporation. He is the author of *CNC for Machining* and *CNC Advanced Techniques*, and is a regular contributor to *Modern Machine Shop* magazine.